Advances in Information Security

Volume 108

The purpose of the *Advances in Information Security* book series is to establish the state of the art and set the course for future research in information security. The scope of this series includes not only all aspects of computer, network security, and cryptography, but related areas, such as fault tolerance and software assurance. The series serves as a central source of reference for information security research and developments. The series aims to publish thorough and cohesive overviews on specific topics in Information Security, as well as works that are larger in scope than survey articles and that will contain more detailed background information. The series also provides a single point of coverage of advanced and timely topics and a forum for topics that may not have reached a level of maturity to warrant a comprehensive textbook.

Javier Hernandez Fernandez • Aymen Omri •
Roberto Di Pietro

Physical Layer Security in Power Line Communications

Fundamentals, Models and Applications

Springer

Javier Hernandez Fernandez
Iberdrola Innovation Middle East
Doha Qatar, Qatar

Aymen Omri
Iberdrola Innovation Middle East
Doha Qatar, Qatar

Roberto Di Pietro
CEMSE Division
King Abduallah University of Science and Technology (KAUST)
Thuwal, Saudi Arabia

ISSN 1568-2633 ISSN 2512-2193 (electronic)
Advances in Information Security
ISBN 978-3-031-57351-4 ISBN 978-3-031-57349-1 (eBook)
https://doi.org/10.1007/978-3-031-57349-1

This Springer imprint is published by the registered company Springer Nature Switzerland AG
The registered company address is: Gewerbestrasse 11, 6330 Cham, Switzerland

To our dear families.

Foreword

Power line communications (PLC) has drawn increasing interest in recent years due to its potential to provide wireline communications services using existing power-line infrastructure. A key issue arising in the use of PLC is that power-line infrastructure was not designed originally for communications purposes and, as an open medium, is generally not secured against potential eavesdropping and other security breaches. Physical layer security (PLS) has emerged in recent years as an alternative or companion technology to more traditional methods of communications security, especially for securing networks of low-complexity devices such as those envisioned for many applications of PLC. Consequently, there has been a considerable amount of work on the application of PLS in PLC in recent years, and this book provides, for the first time, a comprehensive treatment of the state-the-art of this field. It also serves as a very useful introduction to PLC itself, with the initial chapters providing a concise review of the state-of-the art in PLC. By examining the entire range of issues arising in this context, the authors provide readers with the tools to understand and potentially develop PLS techniques for application in PLC networks. This book is an excellent resource for those interested in learning about the potential of PLS to help secure PLC networks, and for those who merely want to learn more about PLC technology. Its breadth and depth of coverage, and its expert insights, make it a singular contribution to the field.

Princeton, NJ, USA
December, 2023

H. Vincent Poor
Michael Henry Strater University Professor
Princeton University

Preface

We live in an ever increasingly interconnected and digitally dependent world, where the critical infrastructure that sustains our modern way of life still shows a worrying degree of vulnerability. In this scenario, the power line communications (PLC) technology and associated paradigms have the potential to emerge as a linchpin, serving as the nervous system that facilitates the seamless flow of information between various components of our energy grid. Moreover, given their pervasiveness, their possible usages are yet to be fully discovered. For instance, PLC could be the silver bullet to provide IoT with the long-sought connectivity grail, or being an enabler for the edge cloud.

Whatever the usages will be, already today, in the realm of PLC, given the convergence of power systems and communication technologies, robust security measures are a pre-requisite—though intertwined with challenges and constraints.

We felt the need to produce this book in an attempt to both unravel the complexities and safeguard the confidentiality, integrity, and availability of power line communications. In particular, we intended to not only clarify the fundamental principles of physical layer security in PLC, but also to provide practical insights into mitigating risks and fortifying resilience.

This book is intended for engineers, researchers, and practitioners struggling with the multifaceted dimensions of cybersecurity in our energy infrastructure. The clarity of the language is accompanied by a rigorous treatment of the topics, and findings are supported by extensive simulations, showing the viability of the several offered solutions, also through the exposition of compelling use cases.

We hope our readers will find this book useful and inspiring, enjoying it at least as much as we enjoyed the research journey that has led to its publication.

Doha, Qatar
Doha, Qatar
Jeddah, Saudi Arabia
May, 2024

Javier Hernandez Fernandez
Aymen Omri
Roberto Di Pietro

Acknowledgments

We extend our gratitude to Iberdrola for their support.

Additionally, we acknowledge the valuable assistance of our colleagues Dr. Gabriele Oligeri, Muhammad Irfan, Abdulah Youssef Jarouf, Ayman Al-Kababji, Dr. Imene Mecheter, and Mohammad Khorasani.

Contents

1 Introduction 1
1.1 Power Line Communication 1
1.2 Power Lines, a Vulnerable Environment 2
1.3 About This Book 5

2 An Overview of Power Line Communication Networks 9
2.1 Standardization Environment and Bandplans 9
2.2 Technologies and Protocols 11
2.3 Applications of PLC 16

3 Power Line Communication Channel Characteristics 21
3.1 Transmission Line Theory 21
3.2 PLC Signal Multipath Propagation 22
3.3 Channel State Information 23
3.4 PLC Channel Impulse Response Characteristics 23
3.5 Additive Noise in PLC 28

4 Power Line Communication Security 31
4.1 PLC Issues and Limitations 31
4.2 PLC Security Threats 33
4.3 PLC Security Attacks 39

5 Physical Layer Security in Power Line Communications 41
5.1 Physical Layer Security 41
5.2 Physical Layer Security in PLC 43
5.3 Physical Layer Security Techniques in PLC 45

6 Performance Analysis of Data Transmission Security and Reliability in PLC 51
6.1 System and Adversary Model 51
6.2 Average Secrecy Capacity 53
6.3 Secrecy Outage Probability 55
6.4 Secure and Successful Transmission Probability 57

7 Confidentiality Technique for PLC Networks 65
7.1 PLC Channel Modeling 65
7.2 PLS Key Generation Scheme 68
7.3 Results and Discussion 73

8 Authentication Method for PLC Network 85
8.1 PLC Channel Impulse Response Model 85
8.2 Physical Layer Identification Scheme Description 86
8.3 Simulation Results and Discussion 91

9 PLC Network Integrity Solution 97
9.1 Background on PLC Topology Change Detection Solutions 97
9.2 PLC System and Channel Models 100
9.3 PLC Multipath Channel Model 101
9.4 Algorithmic Overview of the Topology Change Detection System 103
9.5 Performance Analysis and Numerical Results 108

10 PLC Availability Scheme 113
10.1 Denial of Service Attacks in PLC Networks and Jamming Detection Challenges 113
10.2 Adversary Model and Motivation 114
10.3 Methodology 117
10.4 Performance Analysis 120
10.5 Discussion 122

11 Conclusions 125

References 127

Index 141

Acronyms

AMI	Advanced Metering Infrastructure
ARIB	Association of Radio Industries and Businesses
ASC	Average Secrecy Capacity
AWGN	Additive White Gaussian Noise
BB-PLC	Broadband Power Line Communication
BER	Bit-Error Rate
BGR	Bit Generation Rate
BMR	Bit Mismatch Rate
BPSK	Binary Phase Shift Keying
CDR	Connection Detection Rate
CEN	European Committee for Standardization
CENELEC	European Committee for Electrotechnical Standardization
CEPRI	China Electric Power Research Institute
CFR	Channel Frequency Response
CIR	Channel Impulse Response
CNN	Convolutional Neural Network
CSI	Channel State Information
CSRF	Cross-Site Reference Forgery
CSS	Cross-Site Scripting
CTF	Channel Transfer Function
CUSUM	Cumulative Sum Control Chart
DAA	Destination Address Alteration
DAK	Direct Access Key
DCSK	Differential Chaos Shift Keying
DER	Distributed Energy Resource
DES	Data Encryption Standard
DFC	Dynamic Flow Concept
DL	Deep Learning
DNS	Domain Name System
DoS	Denial of Service
DSO	Distribution System Operator

EMC	Electromagnetic Compatibility
EMI	Electromagnetic Interference
EST	Effective Secrecy Throughput
ETSI	European Telecommunications Standards Institute
EU	European Union
EV	Electric Vehicle
FCC	Federal Communications Commission
FDD	Frequency-Division Duplexing
FDR	Frequency Domain Reflectometry
FIR	Finite Impulse Response
FNDP	False Negative Detection Probability
FPDP	False Positive Detection Probability
G.hn	Gigabit Home Networking
G.hn-MIMO	Gigabit Home Networking-Multiple-Input and Multiple-Output
G3-PLC	3rd Generation Power Line Communication
HAN	Home Area Networks
HD-PLC	High Definition-Power Line Communication
HIF	High Impedance Fault
IBFD	In-Band Full-Duplex
IDA	Information Dispersal Algorithm
IDFT	Inverse Discrete Fourier Transform
IDS	Intrusion Detection Systems
IEEE	Institute of Electrical and Electronics Engineers
IoT	Internet of Things
ISP	Internet Service Providers
ITU	International Telecommunication Union
KDR	Key Disagreement Rate
KGR	Key Generation Rate
LF SGPLC	Low Frequency Smart Grid Power Line Communications
LO	Length Overflow
MAC	Medium Access Control
MF SGPLC	Medium Frequency Smart Grid Power Line Communications
MIMO	Multiple-Input and Multiple-Output
NAN	Near-me Area Network
NB-PLC	Narrowband Power Line Communication
NDSO	National Standard Developing Organizations
NEK	Network Encryption Key
NMK	Network Membership Key
NOMA	Non-Orthogonal Multiple Access
OFDM	Orthogonal Frequency Division Multiplexing
PDF	Probability Distribution Function
PER	Packet Error Rate
PHEV	Plug-in Hybrid Electric Vehicle
PKI	Public Key Infrastructure
PL	Power Line

PL ID	Physical Layer Identification
PLC	Power Line Communication
PLM	Power Line Modem
PLS	Physical Layer Security
PNR	Passive Network Reconnaissance
PRIME	Power Line Intelligent Metering Evolution
RC	Rogue Commands
RF	Radio Frequency
RGB	Red Green Blue
RI	Rogue Interloper
RSA	Rivest Shamir Adleman
RSSI	Received Signal Strength Indicators
SC	Secrecy Capacity
SDO	Standard Development Organization
SDP	Successful Detection Probability
SEE	Secrecy Energy Efficiency
SEM	Standard Error of the Mean
SINR	Signal-to-Interference-Plus-Noise Ratio
SNR	Signal-to-Noise Ratio
SOP	Secrecy Outage Probability
SPDG	Successful Path Detection Gain
SPDP	Successful Path Detection Probability
SQLi	Structured Query Language Injection
SR	Secrecy Rate
SSTP	Secure and Successful Transmission Probability
ST	Secrecy Throughput
TDD	Time-Division Duplexing
TDoA	Time Difference of Arrival
TEM	Transverse Electromagnetic
TL	Transmission Line
ToA	Time of Arrival
TSM	Transport Sequence Modification
UF	Unavailable Function
UNB-PLC	Ultra Narrowband Power Line Communication
VCIR	Virtual Channel Impulse Response
WCR	Wiretap Code Rate

Chapter 1
Introduction

Abstract Although Power Line Communication (PLC) technology has a long history, its broader acceptance and deployment have only gained momentum in recent years, primarily due to its applications in industrial settings. For years, various standardization organizations and industrial associations have been working towards making PLC a widely accepted and secure communication technology. Despite its quick adoption, there has not been much focus on the security of PLC, particularly regarding physical layer techniques, with current measures mostly relying on traditional cryptographic principles. This book provides an in-depth analysis of PLC Physical Layer Security (PLS), examining its techniques, features, and limitations. It also introduces novel metrics and reviews existing ones to deliver a comprehensive security performance assessment of PLC networks. This introductory chapter presents an overview of PLS in PLC and outlines the structure and scope of the remaining chapters of the book.

1.1 Power Line Communication

The Power Line Communication (PLC) industry has seen tremendous advances in the past few years. Fueled by the deployment of smart meters and the rise of the Internet of Things (IoT), PLC technology has experienced a significant growth spurt. At the same time, advances in communication technologies have enabled wider and higher-throughput networks, which can support numerous applications beyond metering and monitoring. For those applications that rely on PLC networks, the opportunities for novel uses of such technologies have ushered in a new era where previously unthinkable implementations are becoming increasingly common.

The potential of PLC comes from its ability to leverage existing power lines and wiring infrastructure for data transmission. This makes it cost-effective and reliable compared to other wired or wireless communication systems. Moreover, the introduction of multiple protocols, such as Power Line Intelligent Metering Evolution (PRIME) or G.hn, has made deploying PLC increasingly easier. As a piece of evidence, a 2022 report indicated that the deployment of smart meters in the

J. Hernandez Fernandez et al., *Physical Layer Security in Power Line Communications*, Advances in Information Security 108,
https://doi.org/10.1007/978-3-031-57349-1_1

European Union (EU) has achieved a regionally varied but significant penetration rate, with an average adoption of 51% [2].

As advancements in PLC are making it easier to access electronic devices for communication, data transfer, and automation purposes, new applications of PLC have begun to emerge. Newer PLC applications have become increasingly popular across the board, impacting a wide range of infrastructures, from private residential networks like home automation systems to industrial operations. As the number of applications increases, so does the potential attack surface, providing more opportunities for hackers. The impact of leaving this issue unaddressed could be long-reaching and devastating. It could cause interference in critical infrastructure components, potentially leading to severe damages and impairing ongoing operations. Such disruptions impact private citizens, businesses, and other organizations that rely on these components for their regular functions while eroding confidence in national security systems.

With greater utilization comes an increase in the demand for reliable security systems: As PLC technology continues to become more established, the need for secure communication channels has only grown larger. In turn, engineers and researchers are tasked with incorporating strong safety protocols into each new development. This ensures that devices remain protected, information stays confidential, and users continue to feel at ease when sending data through such agile backbone systems.

As more applications are developed, it is important for engineers and researchers to ensure that devices can securely transfer data and commands by incorporating key security measures in the planning process. Each PLC setup must encompass defense systems specific to its particular environment, as every new application demands unique steps towards ensuring optimum security coverage. Right from the development phase, planners must map out strategies that include established protections such as authentication, data encryption, and integrity verification. In this regard, Physical Layer Security (PLS) techniques can play a key role by complementing these approaches, providing an additional layer of security, and helping to protect against various types of threats.

Therefore, it is necessary to consider the fundamentals of PLS in PLC while creating an updated outlook. Delving into the details of PLS systems can provide crucial insight—including an understanding of their varied performance metrics. In analyzing these elements, we are given the opportunity to assess current solutions and potential areas for improvement more comprehensively. It is through this process that we can identify the most efficient methods and accurately measure their impact on PLC security. With a strong understanding of these fundamentals, engineers have better chances of ensuring secure communications for new and existing applications.

1.2 Power Lines, a Vulnerable Environment

Although PLC offers certain advantages over other communication options, there are a number of vulnerabilities associated with the technology that must be taken

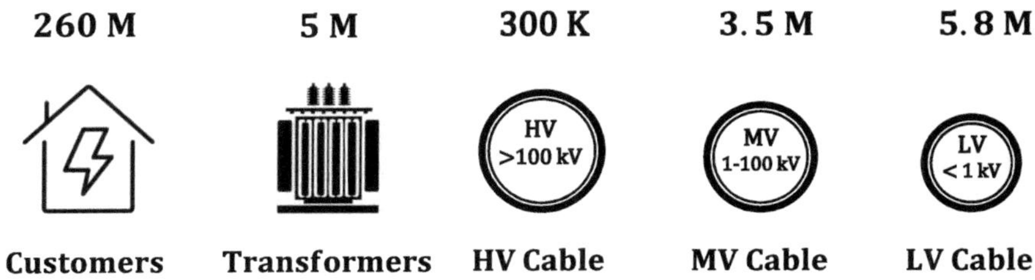

Fig. 1.1 Estimated number of kilometers of cables, transformers, and customers within the distribution network of the EU25 [59]

into consideration, especially as they can disrupt the power grid and damage critical infrastructure.

From a pure telecommunication point of view, PLC systems are vulnerable to electromagnetic interference from the environment, which can affect their signal quality and reliability. Malicious actors may interfere with communications by transmitting their own signal at the same frequency as PLC, leading to jamming. This could be done using devices such as antennas, amplifiers, or any other device that broadcasts noise signals on the PLC frequency. Such an attack can be used to prevent communication between two devices or even manipulate data being sent over these networks. Various security measures need to be implemented to prevent jamming, such as frequency hopping techniques and authentication protocols to ensure secure communications. However, PLC naturally provides a higher level of protection against jamming compared to wireless alternatives.

Another key factor to consider when analyzing the vulnerabilities of PLC is the attack surface. The infrastructure of power cables that distribute energy from generators to homes is extensive, not homogeneous, and largely unmonitored, posing a potential security risk. As shown in Fig. 1.1 and according to data compiled by Eurelectric, a business association representing over 3,500 Distribution System Operator (DSO)s, the EU25, along with Norway, is estimated to control approximately 10 million kilometers of power lines [59]. The study, which does not include the data relative to the power lines operated by the transmission operators, also offers other interesting figures that give an order of magnitude of the size of the attack surface. These mentioned critical infrastructures are susceptible to attacks due to their dispersion, lack of physical security in most cases, and the effects of automation and digitization. The smart grid can enhance power distribution efficiency but also create new opportunities for adversaries to exploit vulnerabilities.

Securing PLC networks is therefore becoming increasingly important, as these devices are often used to transmit sensitive data in various industrial, commercial, and residential applications. However, PLC security remains heavily dependent on cryptographic techniques, with most standards like G.hn or PRIME relying on Public Key Infrastructure (PKI) implementations [250]. In addition, some communication protocols used in PLC networks lack authentication mechanisms, which can lead to external attackers taking advantage of weak passwords or default configurations set by manufacturers. To counter this threat, new standards have also introduced authentication methods designed for use with public keys. For example, PRIME implements digital signatures based on the Rivest Shamir Adleman (RSA) cryptography for secure device identification and authentication [201].

Overall, these measures can be effective at reducing security risks associated with PLC networks. However, further work is needed to develop more robust and comprehensive security mechanisms that can provide end-to-end protection in this type of environment. Techniques that can take advantage of the unique features of the power grid, like PLS, are being explored as potential solutions.

PLS achieves secrecy by utilizing the physical properties of the communication channels, as opposed to relying on cryptographic methods like existing approaches. This represents a unique paradigm that does not rely on the adversary's computation power or having access to any prior information about the network system. The level of security provided by the physical layer is provable and quantifiable, allowing for comparisons between different systems. This can facilitate the evaluation of a system's security and the effectiveness of its physical layer measures. By leveraging the properties of the propagation environment, PLS can exploit physical layer features such as temporal variations, channel reciprocity, multipath fading, or spatial diversity in order to provide an extra layer of security.

Current research efforts are focusing on utilizing Channel State Information (CSI) to develop tailored coding schemes that provide greater control over the security level of transmission. Additionally, exploiting geometry and temporal correlations between PLC modems can enhance the ability to secure transmissions by providing fine-grained information about legitimate endpoints. Furthermore, using machine learning algorithms for CSI training helps create more robust PLS systems that are able to adapt quickly to changes in the environment.

PLS is especially well tailored for resource-constrained devices, such as those found in the IoT given their limitations in terms of power, memory, and computational capabilities. Additionally, PLS can be used in combination with existing cryptographic methods to increase overall security. This combination may be particularly useful when dealing with stronger adversaries that have access to more computing power or prior knowledge about the network system.

The potential applications for PLS are wide-reaching due to its ability to provide robust security with minimal overhead on available resources and that it can be applied to all domains, from wireless networks like WiFi and cellular networks to optical fiber communication systems.

In PLC, PLS can provide a perfect complement to existing cryptographic methods. This is especially important in smart grid systems, given the sensitive nature of their data and the potential for cyber-physical attacks. PLS techniques create a more efficient system by allowing for direct communication between different network elements with minimal energy consumption or latency overhead.

Physical layer techniques have provided creative solutions to communication challenges, especially in wireless mediums, including interference management, synchronization, transmission security, etc. Unfortunately, in PLC, the research in this field is still in its infancy, with few initiatives reported in the literature. A significant obstacle to adopting wireless PLS techniques in PLC is the substantial dissimilarity between the physical media. This is due to the lack of reciprocity in the PLC channel because of the impedance imbalance between the transmitter and the receiver, which is not the case for wireless channels. Hence, many wireless physical layer techniques cannot be adopted directly due to the lack of clear reciprocity in the channel. However, the Channel Impulse Response (CIR) remains positively correlated to the path delays for both cases but not path gains. In this context, this book provides comprehensive information about the topic and suggests different ways to apply PLS in various PLC situations by leveraging the path delays.

This presents an exciting opportunity for further research and development in this field. By integrating PLS techniques into PLC networks, it is possible to create more secure and robust communication systems that are better able to handle cyber threats. The potential applications for PLS in PLC are sizable and, with further research, will continue to expand as new possibilities are identified. There is a need for further investigation into exploiting path delays, multipath fading, temporal variations, or spatial diversity in order to provide an extra layer of security. Moreover, leveraging machine learning algorithms on reciprocal channel characteristics can also be explored as an avenue to create even more robust systems that can adapt quickly to changes in the environment. This will not only improve the overall security of PLC networks but will also ensure more reliable and efficient access to energy for all users.

1.3 About This Book

This book provides a comprehensive examination of PLS in PLC networks. The book is split into two parts, as illustrated in Figs. 1.2 and 1.3. The first part covers the current state of PLC security and its ecosystem, presents new metrics for assessing performance, and discusses existing metrics that are crucial for evaluating the effectiveness of network security solutions. The second part focuses on showcasing how PLS can ensure confidentiality, integrity, availability, and authentication for PLC networks through practical yet rigorous examples.

Chapter 2 provides an outlook of PLC networks, reviewing the various technologies and applications. It also summarizes the main associations supporting this technology, their standards, and their technical capabilities. Furthermore, it explores different applications of PLC, including data transmission and grid information inference. Finally, it covers topics related to channel characteristics, such as transmission line theory, multipath propagation model, and noise. This basic information operates as a powerful starting point for further examinations surrounding PLC security.

The characteristics of the PLC channel are pivotal for assessing the dependability of the systems and linking data transmission performance directly with signal quality. Chapter 3 provides an examination of the transmission line theory, multipath signal propagation, and noise attributes pertinent to PLC networks.

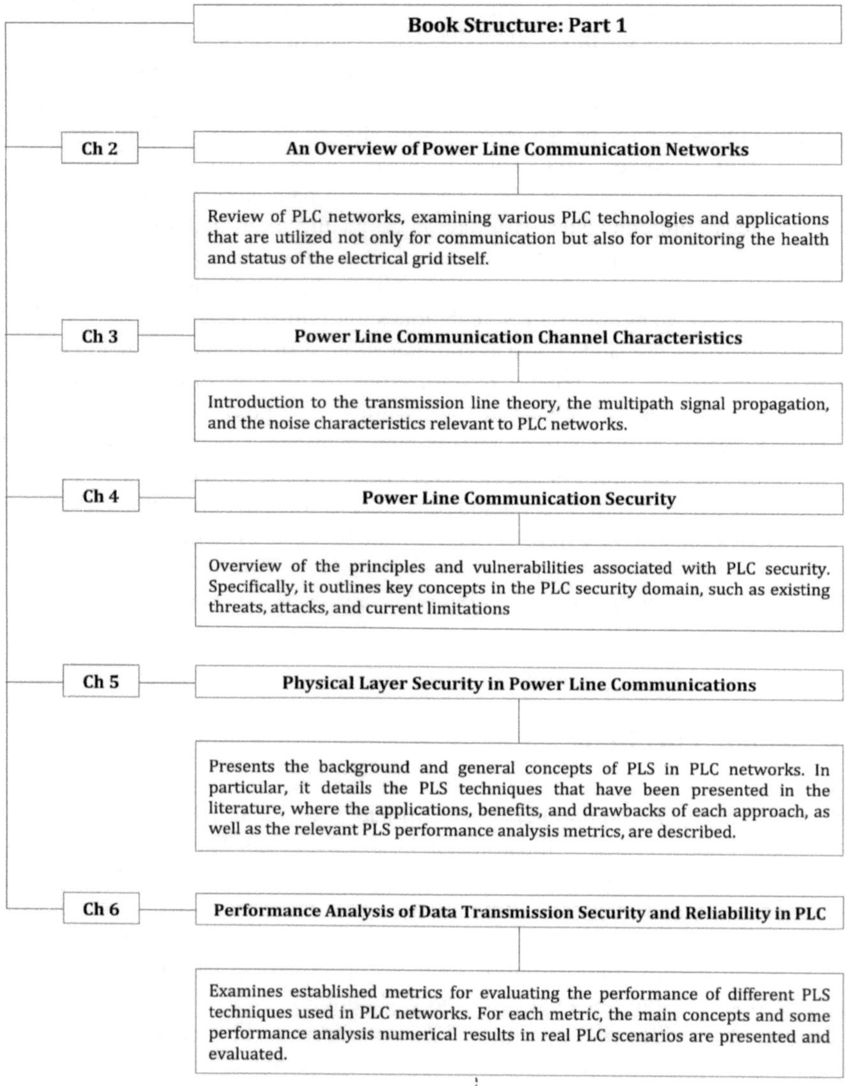

Fig. 1.2 Overview of the structure of the first part of the book

Chapter 4 continues with a description of the PLC security fundamentals and threats. In particular, it introduces the main concepts related to PLC security and details the types of threats the physical layer is subject to. Furthermore, it outlines the security attributes inherent to PLC networks, delves into pertinent security challenges, and addresses the privacy considerations specific to PLC systems.

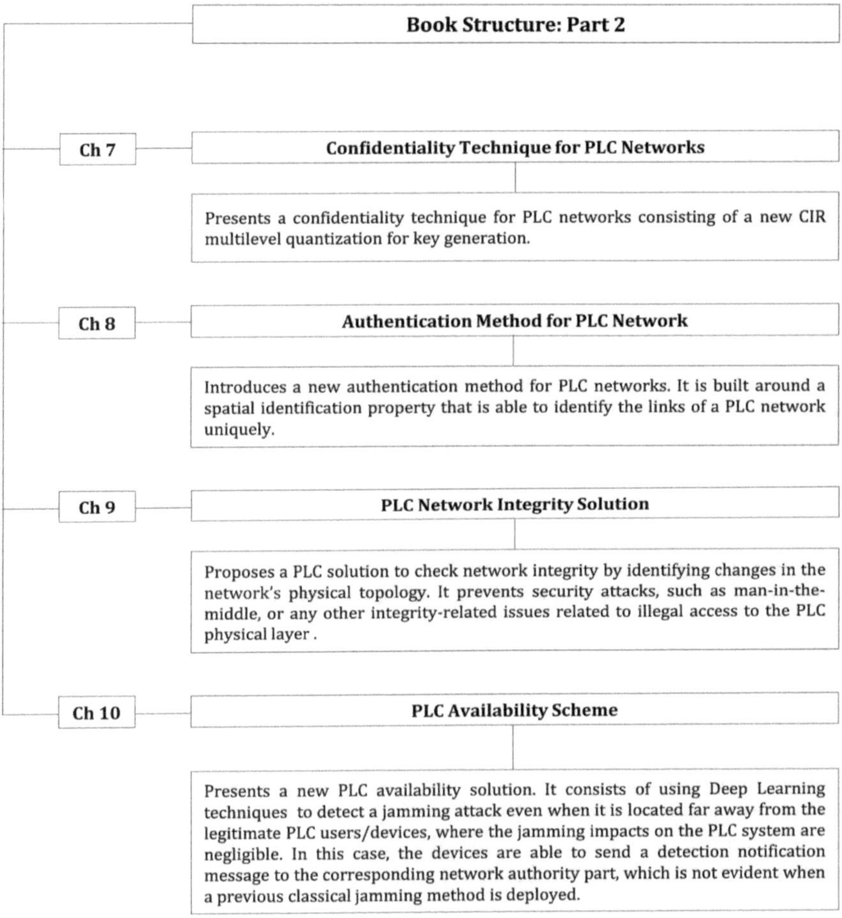

Fig. 1.3 Overview of the structure of the second part of the book

Chapter 5 is devoted to uncovering the general principles of PLS and how they are used in PLC environments. To this end, an in-depth exploration of the various security techniques available to secure the physical layer will be presented. Specifically, these techniques can be classified into three distinct domains: time, spatial, and frequency. An analytical evaluation of each technique's respective benefits and drawbacks will also be discussed. Furthermore, a comparative table outlining the advantages and disadvantages of using each of these categories will provide our readers with an understanding of the best technique applicable to their particular system architecture.

We dedicate Chap. 6 to the performance analysis metrics used in selecting and evaluating PLS techniques for PLC networks. It will assess, specifically, the Secrecy Rate (SR), the Secrecy Outage Probability (SOP), and the Secure and Successful

Transmission Probability (SSTP). For each of these topics, the main concepts and numerical results from using such techniques in real scenarios shall be studied and evaluated.

A confidentiality technique for PLC networks is presented in Chap. 7. It consists of a new multilevel quantization for PLS key generation to prevent eavesdropping. In particular, the new method tackles the asymmetry characterization of power line channels that has made it difficult to use physical layer techniques within PLC by deriving a virtual symmetric CIR from the related asymmetric PLC CSI. Based on that, it will introduce a low-complexity CIR multilevel quantization method that can be deployed in PLC PLS key generation.

Chapter 8 will introduce and detail an authentication method for PLC networks. It starts by introducing a spatial identification technique designed to uniquely identify the links within a PLC network. Initially, the multipath channel characteristics specific to PLC are thoroughly described. Subsequently, a technique for detecting multipath channel delay is presented in order to provide an accurate physical layer identification (PL ID) for the PLC links under consideration. The method's precision is validated through the measurement of the Successful Path Detection Probability (SPDP) within a simulated environment under both perfect and imperfect channel state information conditions. The minimal complexity of the introduced method, along with its promising outcomes, establishes a strong foundation for its application in PLC networks and especially for authentication purposes.

Chapter 9 introduces a PLC solution to check PLC network integrity. This solution consists of a PLC topology change detection technique that can identify changes in the network's physical topology. Experimental results will be presented to validate the advantages of the introduced technique in terms of successful topology change detection accuracy. This software-only-based solution for topology detection presents an efficient PLC network integrity scheme to detect and prevent security attacks, like man-in-the-middle, or any other integrity security issues related to illegal access to the physical layer of PLC.

Chapter 10 introduces a novel solution aimed at enhancing the availability of PLC networks by identifying jamming attacks. Utilizing deep learning methodologies, the proposed technique can detect jamming attempts even when the effects on the PLC modem are minimal. Under such conditions, a device can send an alert before being blocked or communicate on behalf of other devices being targeted. This capability represents a distinctive advantage of the proposed method in comparison to prior jamming detection approaches in the existing literature. Notably, the presented technique can identify a jamming attack when the attacker and the legitimate receiver are separated by a distance of 80 m or less, boasting a detection accuracy of 0.99 and a BER close to zero, thereby challenging traditional methods in detecting such attacks.

The results presented in this book can help improve the security of PLC networks and meet the safety requirements demanded by critical infrastructure applications. Overall, this work aims to serve as a valuable resource for researchers and telecommunication engineers involved in the development and deployment of secure PLC networks.

Chapter 2
An Overview of Power Line Communication Networks

Abstract This chapter provides a comprehensive survey of Power Line Communication (PLC) networks, reviewing the different technologies that shape this field. Additionally, it also explores various PLC implementations with special emphasis on the often-overlooked application of grid information inference. This review is designed to shed light on how PLC can be leveraged not just for communications but also for the extraction and analysis of valuable grid data. Such reflection aims to expand the understanding of the diverse uses of PLC, with a particular focus on enhancing grid management and operations.

2.1 Standardization Environment and Bandplans

Standardization and technical requirements are necessary to guarantee effective and compatible transmissions across power lines. The International Telecommunication Union (ITU) is the principal Standard Development Organization (SDO) that supports the standardization of Power Line Communication (PLC) technology standardization of PLC technology. The European Committee for Standardization (CEN), European Telecommunications Standards Institute (ETSI), and European Committee for Electrotechnical Standardization (CENELEC) are the three major players at the European regional level. The Federal Communications Commission (FCC) in the United States, Association of Radio Industries and Businesses (ARIB) in Japan, and China Electric Power Research Institute (CEPRI) are other active National Standard Developing Organizations (NDSO). Professional associations, some not formally recognized as standardization bodies, can also establish and advance in the development of standards. The Institute of Electrical and Electronics Engineers (IEEE) has been the most significant association in this area, releasing numerous standards to date, the majority of which have generally been embraced by the industry. Additionally, industrial consortia are crucial to the creation, acceptance, and advocacy of these standards.

J. Hernandez Fernandez et al., *Physical Layer Security in Power Line Communications*, Advances in Information Security 108,
https://doi.org/10.1007/978-3-031-57349-1_2

Standardization of PLC is often organized according to the frequency of operation. PLC is usually segmented into three principal categories, which are briefly outlined below and elaborated upon in greater detail in [235]:

- **Ultra Narrowband Power Line Communication (UNB-PLC)**
  The UNB-PLC operates in the low-frequency band within a frequency range of 0.3–3 kHz at a modest data rate of 100 bits per second [17, 187, 189]. Historically, due to its extended transmission range, UNB-PLC has been primarily employed in distribution automation for one-way communications, while two-way communications have been more naturally suited for metering applications.
- **Narrowband Power Line Communication (NB-PLC)**
  The NB-PLC operates over a relatively limited frequency band, from 3 to 500 kHz, depending on the region-specific Electromagnetic Compatibility (EMC) regulations and spectrum assignment policies. In Europe, electromagnetic emissions are regulated under Directive 2014/30/EU [60], with specifications outlined by CENELEC. The EN 50065 standard allocates four bands for specific applications, outlined in Table 2.1 [148]. In the United States, transmission limitations for the 10–490 kHz band are overseen by the FCC under Title 47, Part 15 of the Code of Federal Regulations [68]. In Asia, ARIB in Japan and CEPRI in China allocate similar frequency spectrums, ranging from 10 to 450 kHz and 3 to 500 kHz, respectively [28].
  Regarding data rates, NB-PLC technologies are generally classified into two categories: low data rate, commonly associated with spread-spectrum and single-carrier modalities, and high data rate, which frequently employs Orthogonal Frequency Division Multiplexing (OFDM) modulation schemes. Various application-level protocols can be accommodated by NB-PLC solutions, such as DLMS/COSEM (IEC 613343-1) for metering purposes or LonWorks (IEC 14908-3) and KNX (IEC 14543-3-x) for home and building automation [189]. High-rate OFDM-based NB-PLC technologies like IEEE 1901.2, G.hnem, or PRIME are particularly well-suited for smart grid applications, including demand response, smart metering, public lightning, and grid monitoring [34, 159].
- **Broadband Power Line Communication (BB-PLC)**
  The BB-PLC distinguishes itself by its capability to transmit data at high rates, often reaching several hundred megabits per second (Mbps). It operates over an expansive high-frequency range, spanning from 1.8 MHz up to 250 MHz [227, 238]. This high data throughput and broad frequency spectrum make BB-PLC an ideal candidate for various high-demand applications.
  One notable application is serving as a backhaul solution for Advanced Metering Infrastructure (AMI) [108]. In the realm of smart grids, some AMI deployments might require fast, reliable, and large-scale data transmission to efficiently manage utility services like electricity, water, and gas. The high data rates offered by BB-PLC enable real-time monitoring and immediate action, enhancing the performance and reliability of AMI systems [108]. Additionally, BB-PLC is often employed in Home Area Networks (HAN), where there is a growing need for robust, high-speed connections to facilitate data-intensive tasks. These tasks can

Table 2.1 CENELEC frequency band [250]

CENELEC band	Frequency band [kHz]	Applications
A	3–95	Power utilities' services and applications
B	95–125	All relevant applications and services
C	125–140	In-home networks' applications and services
D	140–148.5	Alarm and security's applications and services

include the streaming of high-definition multimedia content, online gaming, or handling a multitude of smart home devices. The BB-PLC technology has proven its efficacy in HAN by offering high-speed connectivity and lower latency, as explored in various studies [158, 214].

2.2 Technologies and Protocols

To bring technologies in alignment with the standards set forth by SDOs, various groups and alliances are actively involved in designing detailed specifications for distinct PLC technologies and semiconductor components [245]. Below is a compilation detailing the main NB-PLC and BB-PLC technologies presently accessible in the market. The comparative analysis of these technologies is visually represented in Figs. 2.1 and 2.2:

2.2.1 NB-PLC Technologies

2.2.1.1 Power Line Intelligent Metering Evolution (PRIME)

This is a low-cost, open, and global PLC standard designed for real-time management of energy and smart metering applications [197]. The PRIME standard was released by the ITU in February 2013 [111] and has published two versions. The initial specification (version 1.3.6), targeting the CENELEC A band, was released in February 2008 and detailed the physical, Medium Access Control (MAC), and convergence layers [111, 199]. In November 2014, released version 1.4 was published in order to incorporate feedback from the field and adjust it to new markets and regulations [201]. The key modifications to the specification were a bandwidth extension to 471 kHz [200] and two resilient modes for enhanced communication in challenging environments. Advanced Metering Infrastructure (AMI), monitoring and control, demand response, microgrid management, and tele-control applications are common uses for PRIME [233, 234]. Table 2.2 lists the PRIME technology's key characteristics in contrast to other PLC systems.

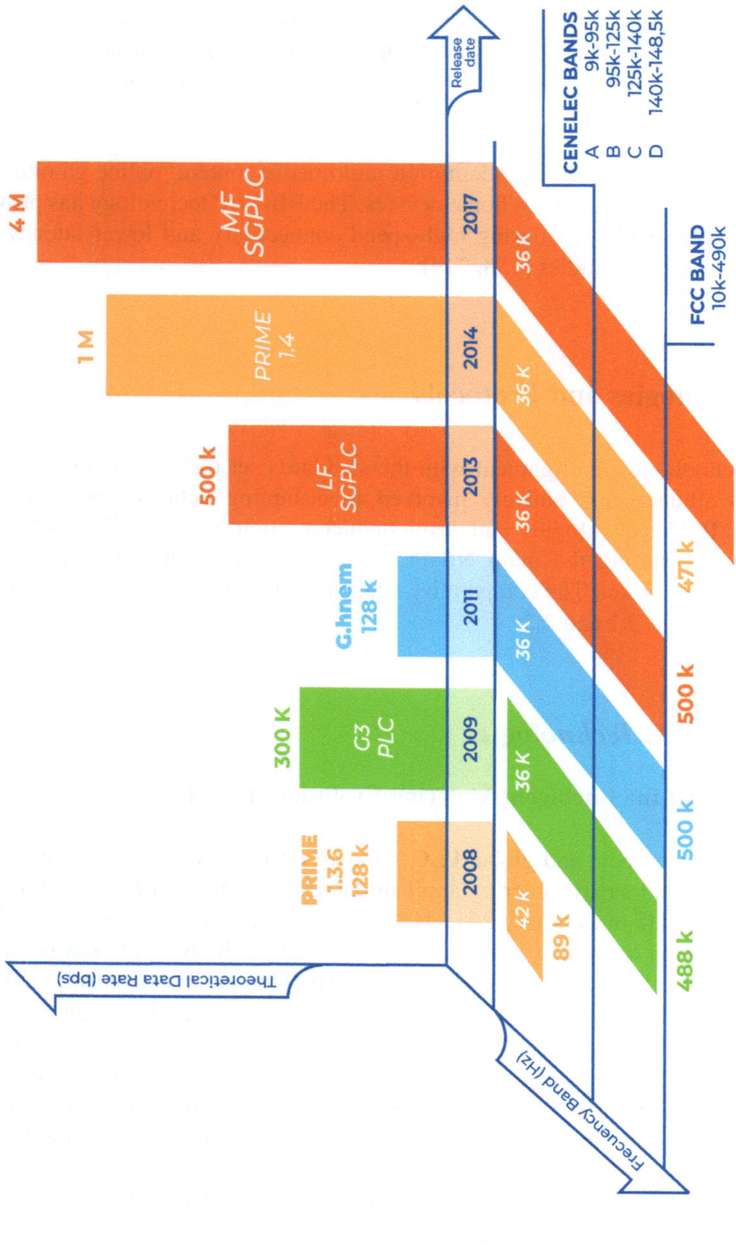

Fig. 2.1 NB-PLC technologies: frequency bands and corresponding theoretical received data throughput [250]

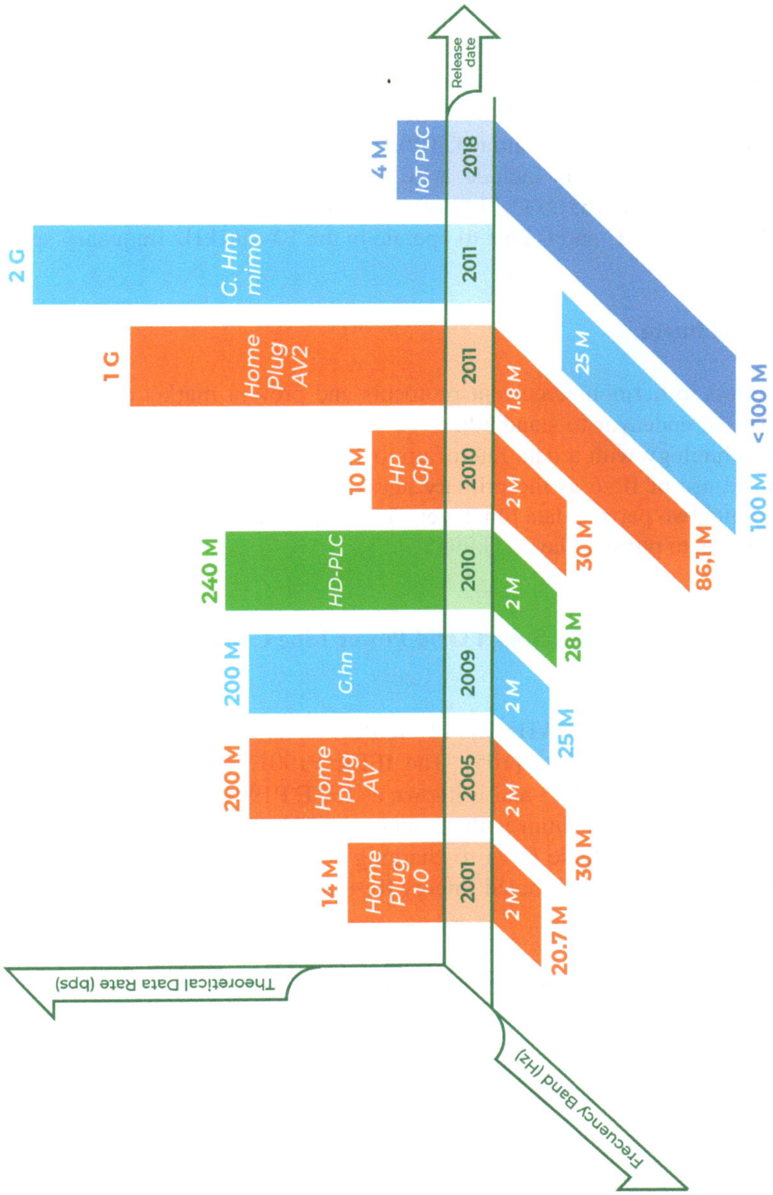

Fig. 2.2 BB-PLC technologies: frequency bands and corresponding theoretical received data throughput [250]

2.2.1.2 3rd Generation Power Line Communication (G3-PLC)

G3-PLC was developed to provide an affordable, long-range, and dependable telecommunication technology over the current power line infrastructure [65]. In July 2009, ITU issued the initial G3-PLC specifications as a standard [22, 114]. Some of its applications include AMI, demand response, load balancing, and prepayment systems [65]. G3-PLC offers a data rate of roughly 300 and 46 kbps in CENELEC-A [21] and is able to operate in the 10–490 kHz frequency spectrum.

2.2.1.3 G.hnem

In an effort to harmonize current protocols and prevent market division, the ITU launched an endeavor to standardize NB-PLC [172]. It outlines a unified OFDM-based technology with a data rate of up to 128 kbps for versatile grid applications that make use of IPv6 as the primary networking protocol [112]. The standard is divided into two parts: a data link layer, G.9956 [112], and an ITU recommendation for the G.hnem physical layer, G.9955 [113].

2.2.1.4 Low Frequency Smart Grid Power Line Communications (LF SGPLC)

At frequencies below 500 kHz, the LF SGPLC technology enables communication for smart grid applications [105]. The IEEE P1901.2 LF SGPLC standard was released in December 2013 and is known as IEEE P1901.2. This standard provides support for low and medium voltage, and it is designed to cover several domains, including Electric Vehicle (EV) to charging stations, grid to smart meters, HAN, and street lighting or Distributed Energy Resource (DER) control.

2.2.1.5 Medium Frequency Smart Grid Power Line Communications (MF SGPLC)

The MF SGPLC IEEE standard provides a PLC solution specifically tailored for smart grid applications. Operating in the medium-frequency range below 12 MHz, this standard offers data transmission rates of up to 1 Mbps [106]. Initially introduced as IEEE P1901.1 in January 2017, it received formal approval in 2018.

2.2.2 BB-PLC Technologies

2.2.2.1 Gigabit Home Networking (G.hn)

It is created by the ITU in the context of home networks, covering power lines, cables and several transmission channels including Cat-5, coaxial (TV), and twisted pair (telephone) cables [171]. The 2–25-MHz frequency spectrum is used by G.hn technology to provide a data rate of up to 200 Mbps for PLC. It has two components: one for the physical layer, ITU recommendation G.9960, which was released in March 2010 [115], and the other for the data link layer, ITU recommendation G.9961, which was published in July 2011 [116]. The PRIME Alliance has standardized the first use case for broadband power line (BPL), the BPL Smart Meter Gateway solution, based on the ITU's G.hn standard [46]. This solution delivers cost-effective load management and security capabilities that enable stability and operational reliability in the grid [198, 221].

2.2.2.2 Gigabit Home Networking-Multiple-Input and Multiple-Output (G.hn-MIMO)

G.hn-MIMO is an extension of G.hn for those cases where three electrical phases are available [34] or where protective earth cables may be utilized [81] to supplement two-wire transmission. The extension uses the MIMO technique to boost the data throughput up to 2 Gbps. ITU released the G.hn-MIMO recommendation, ITU G.9963 [117], in July 2012.

2.2.2.3 High Definition-Power Line Communication (HD-PLC)

It is a technical specification based on the IEEE P1901 standard and maintained by the HD-PLC Alliance [107]. It operates in the 2–28-MHz frequency band to attain optimum transfer speeds of 240 Mbs for home area networks, smart appliances, and other consumer electronics items [83].

2.2.2.4 IEEE P1901.3

Published in December 2018, this IEEE standard employs wavelet OFDM and outlines specifications for both the physical and MAC layers. This technology is specifically geared towards Internet of Things (IoT) applications and is currently recognized as IoT PLC [104]. Panasonic, the founding member of the HD-PLC Alliance, declared in 2018 their intent to adopt IEEE P1901.3 as their fourth-generation HD-PLC technology. They also committed to ensuring backward compatibility with their preceding implementations [181].

2.2.2.5 HomePlug:

The term "HomePlug" encompasses a range of PLC technologies designed for networking through preexisting residential electrical wiring. While some HomePlug standards are tailored for broadband applications–such as broadcasting low data rate IPTV, gaming, and web-based content within a home–others are engineered for lower power, reduced throughput, and more extensive operating temperature ranges to cater to use cases like smart energy meters and intra-home communications among electrical systems and devices. These specifications originate from the HomePlug Powerline Alliance, which also holds the rights to the HomePlug trademark. A detailed breakdown of the various technical specifications across HomePlug versions can be found in Table 2.2.

2.3 Applications of PLC

To accommodate emerging innovations such as DER and electric vehicles (EV)/plug-in hybrid electric vehicles (PHEV), the existing electric grid is undergoing continuous modernization [18, 96, 121, 207]. A reliable communication network is essential for the smart grid architecture to support data transfers and bidirectional communications [66]. PLC not only fulfills this requirement but also allows for the extraction and inference of underlying grid information. As a result, smart grid applications of PLC can be broadly categorized into two main clusters: (i) data transmission and (ii) grid information inference.

2.3.1 PLC for Data Transmission

PLC serves as an effective communication medium across the different segments of the electrical grid, endorsing a range of applications with its reliable and efficient deployment capabilities. Notably, PLC has been instrumental in advancing AMI by equipping a preexisting network for data transmission. Furthermore, the technology has streamlined home automation and appliance control, as the use of PLC eliminates the need for additional control units and complex wiring schemes for signal transmission [256]. BB-PLC networks offer utilities an avenue to provide Internet services to end users and can be seamlessly integrated with other forms of communication networks, such as Ethernet and wireless systems, facilitating both inter- and intra-network data exchanges [29, 145, 163, 218]. The versatility of PLC extends to automotive applications, where it has been examined for in-vehicle communication and control systems [12, 50, 190].

Table 2.2 NB and BB PLC technologies [250]

Category	Technology	Standard/specification	Date of first specification release	Band (Hz)	Peak data rate (bps)	Modulation type	Coding type
NB-PLC	PRIME	ITU G.9904 [111]/PRIME 1.3.6 [22, 199]	February 2008	42–89 k	128 k	DBPSK, DQPSK, or D8PSK	CC
		ITU G.9904 / PRIME 1.4 [201]	November 2014	42–471 k	1 M	DBPSK, DQPSK, or D8PSK	CC and RC
	G3-PLC	ITU G.9903 [114, 222]	July 2009	36–91 k	46 k	DBPSK, DQPSK, or D8PSK	CC, RC and RSC
				10–490 k	300 k		
	G.hnem	ITU G.9955 For PL [112]	February 2011	3–500 k	128 k	DBPSK, DQPSK, or D8PSK	CC, RC and RSC
		ITU G.9956 For DLL [113]					
	LF SGPLC	IEEE P1901.2 [105]	December 2013	9 to 500 k	500 k	DBPSK, DQPSK, or D8PSK	CC, RC and RSC
	MF SGPLC	IEEE P1901.1 [106]	January 2017	< 12 M	4 M	DBPSK, DQPSK, or D8PSK	CC, RC and RSC
BB-PLC	G.hn	ITU G.9960 for PL and the architecture [115]	October 2009	2–25 M	200 M	Up to 4096-QAM	LDPC
		ITU G.9961 For DLL [116]					
	G.hn-MIMO	ITU G.9963 [117]	December 2011	25–100 M	2 G	Up to 4096-QAM	RC
	HD-PLC	IEEE 1901 [107]	January 2010	2–28 M	240 M	2-PAM to 32-PAM	CC, RC, RSC, TCC, and LDPC
	IoT PLC	IEEE P1901.3 [104]	December 2018	<100 M	4 M	BPSK, QPSK, 16-QAM	CC, RC and RSC
	HomePlug	HomePlug 1.0 [91]	June 2001	4.49–20.7 M	14 M	BPSK, DBPSK or DQPSK	CC,RC, RSC and TCC
		HomePlug AV [92]	August 2005	2–30 M	200 M	BPSK, QPSK, 16/64/256/1024-QAM	TCC
		HomePlug GP [93]	June 2010	2–30 M	10 M	QPSK only	TCC
		HomePlug AV2 [94]	December 2011	1.8–86.13 M	1 G	Up to 4096-QAM	TCC

2.3.2 PLC for Grid Information Inference

The utility of PLC transcends mere telecommunication roles and can serve broader functions within the electrical grid itself [223]. Leveraging power lines, Power Line Modem (PLM)s can function as sensing devices by emitting various types of signals across multiple frequency ranges [215]. Subsequent data analysis collected from these PLMs can yield insights into the grid's physical layer attributes. Such applications, herein referred to as "PLC grid information inference," take advantage of PLC's inherent "through the grid" capability for purposes like grid monitoring [66]. Additionally, the cybersecurity domain utilizes both grid and channel attributes to strengthen security and privacy measures. An exhaustive classification and analysis of these application domains are articulated in subsequent sections and visually summarized in Fig. 2.3.

2.3.2.1 Grid Monitoring

Owing to changes in the grid, such as line impedance, loads, and topology, the PLC channel is time-variant [150]. Deviations in grid parameters may also affect other grid activities, including asset management, communication, or fault identification [134, 138]. Therefore, ongoing grid surveillance is necessary to ensure prompt responses to the shifts in grid dynamics [206]. The identification of anomalies and topology inference are two of the key uses of grid monitoring.

Topology inference refers to the ability to deduce the physical arrangement of power line networks based on observations of the current state of the grid. The most frequently examined parameters include the number of nodes, location, and distances between them. This subsection categorizes topology inference based on (1) the degree of existing topological information, (2) the inferred topology parameters, and (3) the frequency of estimation. The three basic subtypes of topological inference we identified are topology estimation, reconstruction, and mapping.

Topology estimation involves predicting the power line network's configuration, taking into account factors such as the number of nodes, interconnections, distances, and the spatial organization of nodes [183, 257]. Topology reconstruction uses the prior knowledge of the topology estimation to recover some or all of the topological parameters [14, 240]. Finally, topology mapping assigns each component to specific grid domains and use cases [52, 182]. The scope of topology estimation can be extended to include the identification of alterations in the network, such as the addition or removal of nodes. Therefore, topology change detection can be viewed as the dynamic counterpart of topology estimation, continually monitoring network variations and issuing alerts for potential intrusions.

Conversely, anomaly detection involves recognizing, categorizing, and locating atypical conditions within the electrical grid. Such anomalies incorporate a range of undesirable events, including line faults, damaged conductors, and node mal-

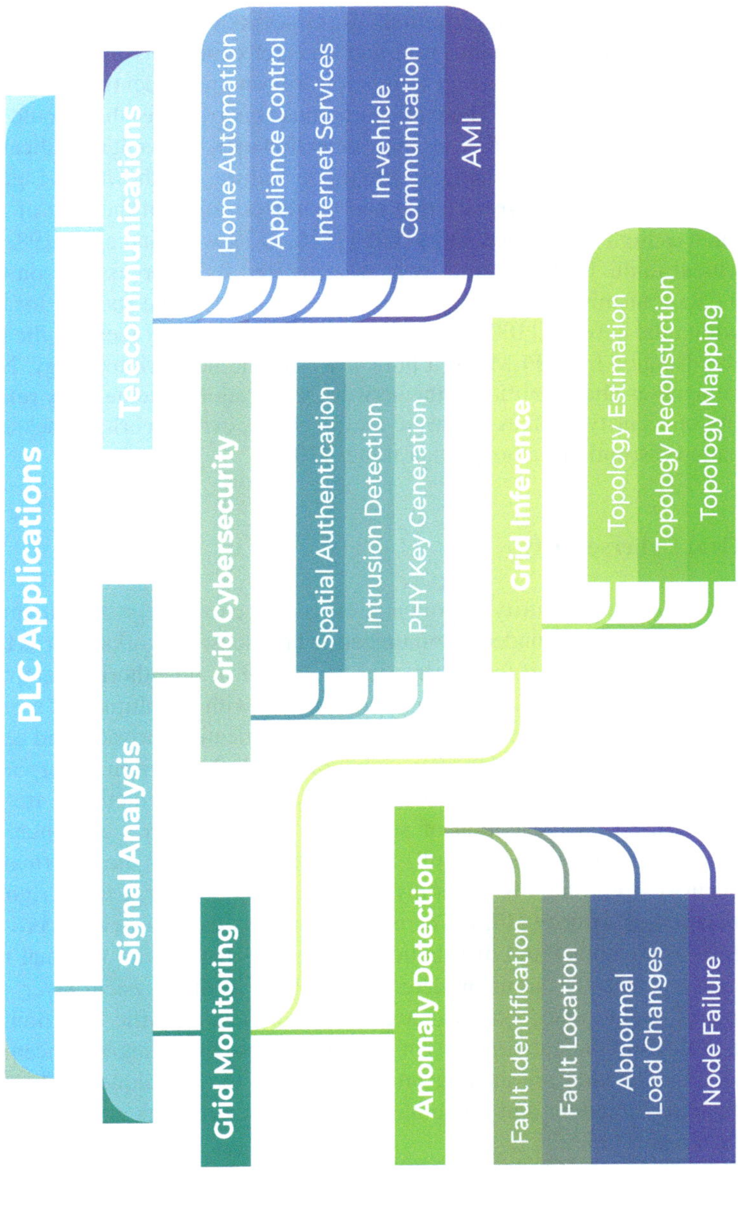

Fig. 2.3 Classification of PLC applications [118]

functions. Notable limitations of contemporary cable diagnostic techniques include specialized hardware requirements, the necessity of conducting tests offline, and the operational complexity involved [100]. Therefore, a straightforward digital processing technique that does not interfere with regular grid activity is a crucial component of effective smart grid anomaly recognition. The concerns previously identified can be resolved by constantly reviewing grid and channel conditions using PLMs to detect anomalies. Most of the existing efforts towards anomaly identification focus solely on recognizing line faults. There are numerous abnormalities that should also be taken into account. We categorize anomaly detection uses into fault identification, fault location, abnormal load changes, and node failure, as shown in Fig. 2.3. Fault identification entails the recognition and characterization of the type of fault involved, such as high-impedance faults (HIF) [156, 157], wear and aging [98, 99], or more critical faults [185], as well as its severity and its repercussions on both power and data transmission. Fault location serves to identify the specific location of a fault within the network [100, 184]. The point of reference for locating the fault could be the distance from PLMs or a branch within the network topology. Node failure encompasses abnormalities in the network's operation, whether they refer to functionality or security, such as hardware malfunctions, cybersecurity attacks, and communication disruptions, among other issues [203].

2.3.2.2 Grid Cybersecurity

The significance of grid security is growing, especially as smart grid deployments progressively integrate unbounded communication broadcast networks such as PLC. This openness paves the way for adversaries to engage in unauthorized activities [72]. Safeguarding smart grids necessitates compliance with a multitude of security prerequisites, including confidentiality, integrity, authentication, privacy, and access control. While cryptographic algorithms and protocols have been incorporated into PLC designs to provide certain degrees of security, these methods are fundamentally premised on the notion of surpassing the adversary's computational capabilities and resources. As a result, the implementation of the information-theoretic security algorithms is generally favored, as they offer a more rigorous conceptualization of security [193]. Therefore, there is a need for new, secure, and streamlined key agreement and distribution approaches that require minimal node interactions. Concurrently, intrusion detection methods at the physical layer, such as detection of unauthorized physical access and malicious traffic injection, are crucial for mitigating the vulnerability of network nodes. Emerging advancements like spatial authentication are gaining prominence due to the unique attributes of node connectivity in indoor and home-based power line networks and the specific noise characteristics [133, 253].

Chapter 3
Power Line Communication Channel Characteristics

Abstract The Power Line Communication (PLC) channel characteristics are key in determining the reliability of PLC systems, as they establish a direct connection between data transmission and signal quality. This chapter offers a comprehensive overview of the transmission line theory, the multipath signal propagation, and the noise characteristics relevant to PLC networks.

3.1 Transmission Line Theory

Power lines serve as the backbone infrastructure of the electrical grid, and a thorough understanding of their characteristics is critical for effective power delivery, loss mitigation, and the reliability of communications. Channel estimation often utilizes transmission line modeling and theory [154, 261], where transmission lines are typically modeled as two-port conductors: one port connects to the power source and the other to the load. The concept of parallel wire configuration involves dissecting a transmission line of length z into an infinite cascade of equal-length segments Δz. The lumped-element circuit model is widely adopted for Transverse Electromagnetic (TEM) lines, where each segment Δz consists of series resistance $R^{'}$ and inductance $L^{'}$, as well as parallel capacitance $C^{'}$ and conductance $G^{'}$ elements, collectively referred to as transmission line parameters, as depicted in Fig. 3.1 [232].

The characteristic impedance Z_0 stands as a key attribute of transmission lines, and its value is shaped by the geometry and materials constituting the line. In the context of a uniform line, Z_0 is invariant with respect to the length of the line. Utilizing the transmission line model founded on telegrapher's equations [232], the general formula for determining the characteristic impedance of a transmission line is as follows:

$$Z_0 = \sqrt{\frac{R^{'} + jwL^{'}}{G^{'} + jwC^{'}}}, \tag{3.1}$$

J. Hernandez Fernandez et al., *Physical Layer Security in Power Line Communications*, Advances in Information Security 108,
https://doi.org/10.1007/978-3-031-57349-1_3

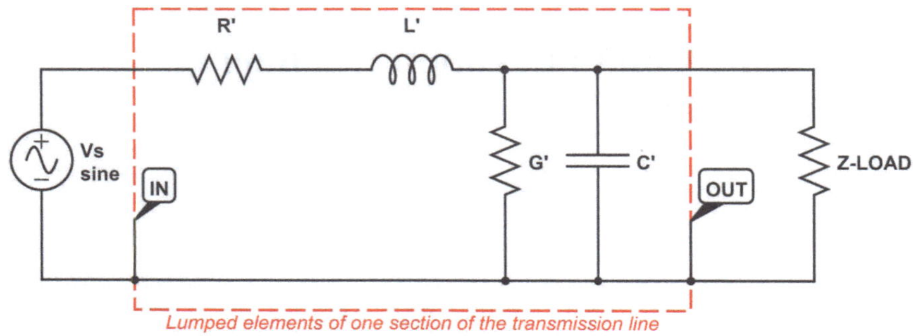

Fig. 3.1 Transmission line model [118]

where w represents the angular frequency measured in radians per second. It should be noted, however, that characteristic impedance is an empirically determined quantity; its value is defined as the ratio of the voltage amplitude V_i to the current amplitude I_i of a wave traveling along the transmission line. For an infinite transmission line, the input impedance aligns with the characteristic impedance, given that no wave reflections occur at the line's end. Discrepancies in impedance values across different segments of the line can induce discontinuities, leading to distortions and reflections in the signal propagation. Such impedance mismatches at nodes, branching points, and line terminations have a consequential effect on the amount of signal reflection or propagation. These phenomena of signal reflection and propagation are critically influential in channel estimation, particularly for applications aimed at inferring grid information in Power Line Communication (PLC) systems.

3.2 PLC Signal Multipath Propagation

Signal multipath propagation in PLC occurs due to signal reflections at branching points, cable joints, and terminations, with impedance mismatches leading to the creation of new paths spreading in various directions [144]. As each signal propagation path is encountered by more reflections, the number of directional paths will increase exponentially, only limited by the significant signal transmission and reflection attenuation. Therefore, a set of delayed and attenuated signal echoes will be received by the corresponding communication devices.

This particular mode of signal propagation, characterized by the signal traversing various paths and encountering varying time delays for each of its replicas, is referred to as multipath propagation [263].

3.3 Channel State Information

Signal propagation within a communication channel is susceptible to alterations arising from the characteristics of the transmission medium itself. Such alterations—which may include factors like fading, scattering, noise, interference, and multipath propagation—directly impact the quality and reliability of the transmitted signal [73]. Broadly speaking, communication channels are categorized into two primary types: fast fading and slow fading channels. Fast fading refers to the rapid fluctuations in channel conditions affecting individual transmitted symbols, whereas slow fading pertains to the more stable or gradually varying channel conditions. To enhance signal reception quality and ensure reliable communications at high data rates, these variable channel conditions must be accounted for through channel estimation and equalization techniques. For slow-fading channels, instantaneous channel estimation methods are generally appropriate, while statistical techniques are more suitable for fast-fading channels. Regardless of the specific estimation methodology employed, the channel state—often abbreviated as Channel Transfer Function (CTF) —is typically represented either by the Channel Frequency Response (CFR) or Channel Impulse Response (CIR), denoted by $h(t)$ and $H(f)$, respectively. The relationship between $h(t)$ and $H(f)$ is governed by the Fourier transform equation:

$$H(f) = \int_0^{\infty} h(t)e^{-j2\pi ft}dt. \tag{3.2}$$

The CFR can be derived from the multipath propagation phenomenon by summing the reflection and transmission coefficients for each individual path, as indicated in Eq. (3.3). Assuming that K paths are identified and each path contains M points of discontinuity, the CFR can be formulated as follows:

$$H(f) = \sum_{k=1}^{K} \left(\prod_{m=1}^{M} T_k^m \right) \left(\prod_{m=1}^{M} \Gamma_k^m \right) e^{-d_k \gamma(f)}. \tag{3.3}$$

In this equation, Γ_k^m and T_k^m represent the reflection and transmission coefficients at the m discontinuity along the k path, respectively. Additionally, d_k specifies the length of the k path.

3.4 PLC Channel Impulse Response Characteristics

The introduced work in this book has taken advantage of the fact that the times of arrival of the various paths between a given transmitter and a given receiver can be collected to form a useful path delay profile.

Accordingly, the following subsection examines the main features of the PLC path delay profiles and how changes in the relevant power line topology affect them.

3.4.1 Reciprocity in PLC CIR

Although PLC CFR is not reciprocal, it has been proven that PLC CIR presents reciprocity in terms of path delays [186]. This wide-sense symmetry of topology-invariant channel analysis is presented and investigated in [186] and shows that the PLC CIRs of two corresponding channels between two given PLC nodes are not strictly symmetric but are wide-sense symmetric. This implies that the uplink and downlink PLC channels between two nodes present the same positions of the corresponding CIR peaks, which is not the same for the relevant peaks' amplitudes; thus, the PLC channel is not strictly symmetric.

To further describe this characteristic, an example of the normalized CFRs and CIRs uplink and downlink of a PLC channels is presented in Fig. 3.2.

As shown in this figure, although a certain similarity between the CFRs can be observed, it is clear that the CFR is not symmetric. On the other hand, the path delays of the CIRs demonstrate a clear symmetry, which is not the case for the amplitudes of the CIRs' peaks. This amplitude's mismatch can be explained by three main reasons: (i) the PLC channel is suffering from a high noise level, in particular, impulsive noise; (ii) a low accuracy of the deployed peak detection algorithm; and (iii) the signal sampling time, which has a significant impact on the CIR's peak detection capability.

In the following section, the topology effects on path delays are investigated.

3.4.2 The Influence of Topology on Path Delays

In the literature, many studies have addressed the topic of the power line topology influence on PLC [37, 38, 77–79]. The authors of these works have investigated the impacts of cable lengths, branches, and impedances on the CIR and the CFR of different PLC topologies.

To further investigate this point, we consider an example of a standard T network structure that consists of two edge nodes A and B, with a middle branch connected to a termination point D as shown in Fig. 3.3, where $|h(t)|$ denotes the CIR magnitude, and τ_i is the corresponding path delay index i, $i \in \{0, 1, ..., R\}$.

By considering the most common topology changes, the theoretical behavior of the relevant multipath delays can be summarized as follows:

- **Transmission distances:** Increasing the transmission distance results in an increase in all multipath delays. On the contrary, decreasing this distance equally

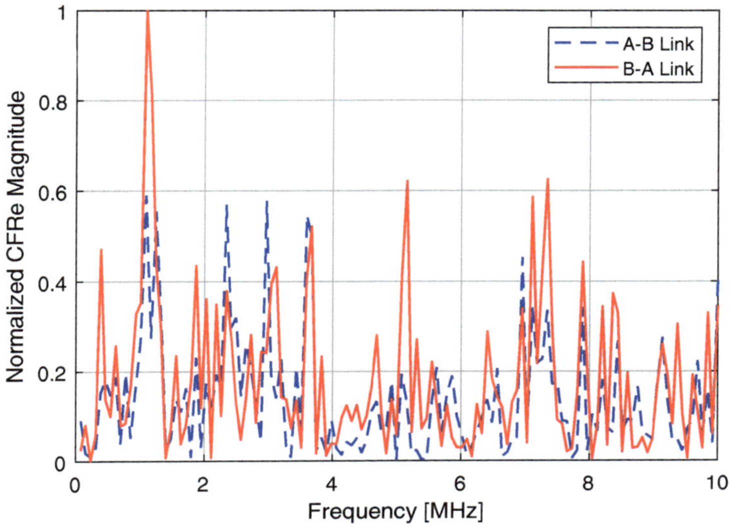

(a)

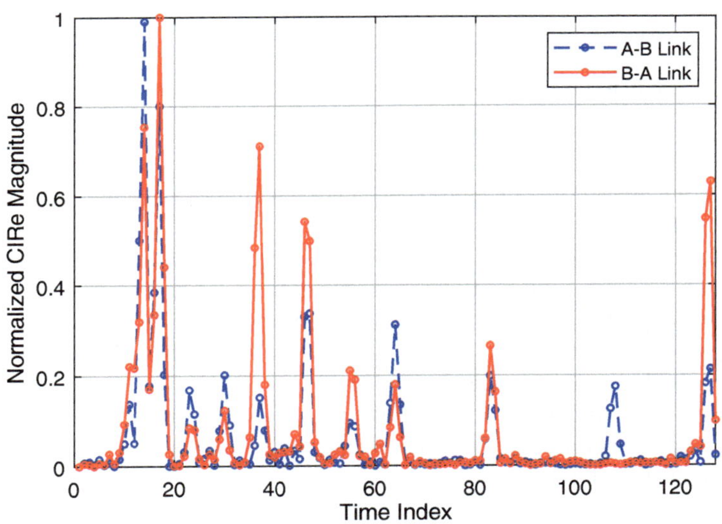

(b)

Fig. 3.2 Example of estimated (**a**) CFRs and (**b**) CIRs of $A - B$ and $B - A$ links [88]

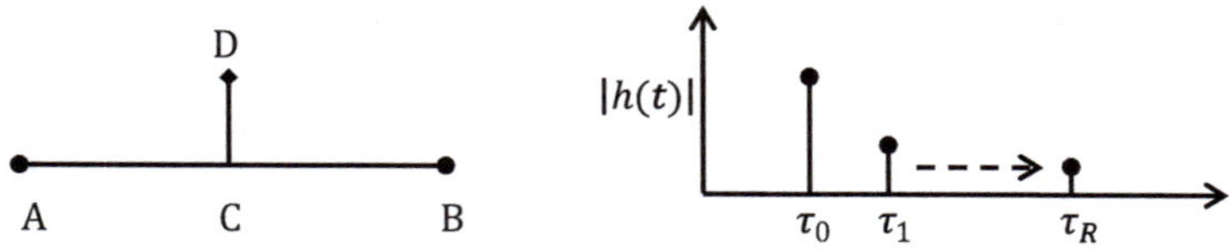

Fig. 3.3 A CIR of the link between two endpoints A and B, in a typical T topology [88]

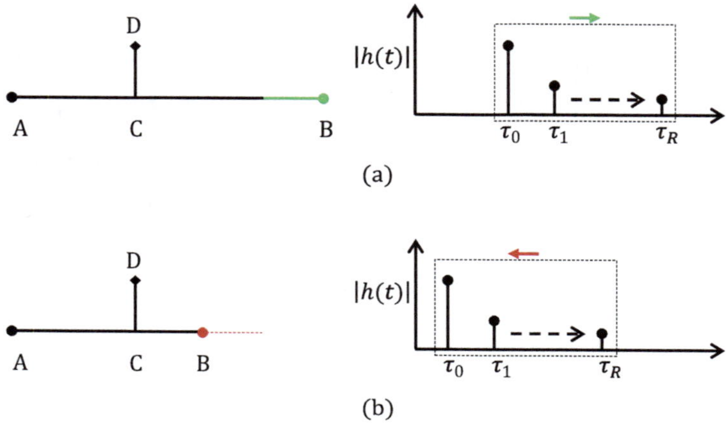

Fig. 3.4 CIR for (**a**) increase and (**b**) decrease in distance between the transmitter and the receiver [88]

reduces the arrival time of each path. These mentioned behaviors are illustrated in Fig. 3.4.

- **Length of branches:** As illustrated in Fig. 3.5, an increase or decrease in the length of a branch between two communicating nodes has no effect on the first path delay τ_0. However, this length change results in a shift of all the remaining multipath components, i.e., a delay in the case of the branch extension or an advance in the case of branch shortening.
- **Number of branches:** The number of received impulses can be increased (decreased) by adding (removing) branches to the same node or along the power line, as illustrated in Fig. 3.6.

It is worth noting that impedance change has no direct effect on the PLC multipath channel delays. However, as it impacts the signals' attenuation, it could affect the significant path delay profile if the signal levels fall below the receiver's sensitivity. However, this phenomenon will not affect the PLC channel reciprocity in terms of path delays since this loss will simultaneously occur on both the transmitter and receiver sides.

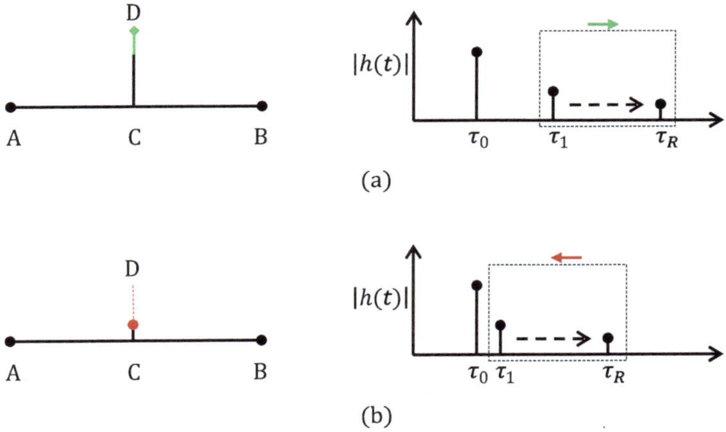

Fig. 3.5 Response of path delay impulses for (**a**) increase and (**b**) decrease in the branch length [88]

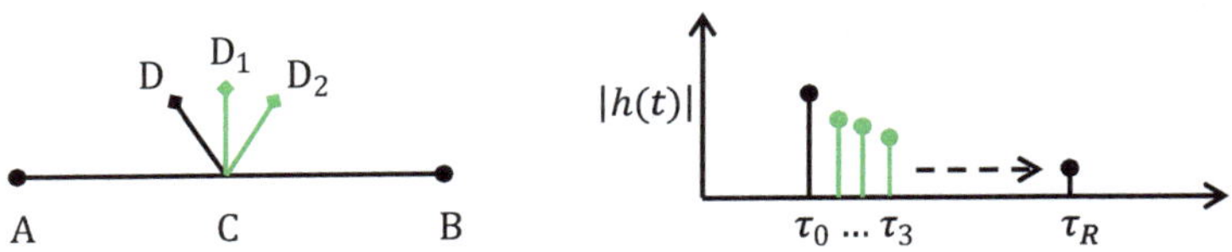

Fig. 3.6 Response of path delay impulses with the increase in branching points [88]

Fig. 3.7 A simple daisy chain connection with three nodes [88]

3.4.3 Path Delay Detection Resolution

As the associated CIR identifies the PLC channel multipath delays, its accuracy highly affects the path delay detection resolution. This accuracy is mainly affected by the sampling frequency used. In fact, increasing the sampling frequency enhances the possibility of detecting all the available paths, even the ones generated by near nodes, and hence, the path delay detection resolution can be improved.

To further clarify this point, a considered scenario is presented in Fig. 3.7, where a power line cable connects nodes A, B, and E in a daisy chain. In this illustration, d_{AB} denotes the distance between A and B, and d_{BE} denotes the distance between B and E, with $d_{AB} < d_{BE}$. In addition, let F_{samp} denote the sampling frequency and $v_p(= 0.6\ C)$ denote the power line signal propagation speed, with C as the speed of light and d as the detection resolution in meters.

In this context, the path delay detection resolution is defined as the minimum distance between two detectable adjacent nodes, e.g., B and E.

Accordingly, the PLC signal propagation times for the direct path ($A \rightarrow B$) and the second path ($A \rightarrow B \rightarrow E \rightarrow B$) are expressed as $t_0 = d_{AB}/v_p$ and $t_1 = (d_{AB} + 2\, d_{BE})/v_p$, respectively. In this case, the minimum sampling time of $T_{samp} \leq (t_1 - t_0)$ is required to detect both mentioned paths at node B. In the case of $T_{samp} > (t_1 - t_0)$, the second path cannot be detected. Hence, the F_{samp} must satisfy the following condition:

$$\begin{aligned} \frac{1}{F_{samp}} &\leq (t_1 - t_0) \\ &\leq \frac{2\, d_{BE}}{v_p}, \end{aligned} \tag{3.4}$$

which can be represented as follows:

$$\begin{aligned} d_{BE} &\geq \frac{v_p}{2\, F_{samp}} \\ &\geq \frac{0.3\, C}{F_{samp}}. \end{aligned} \tag{3.5}$$

Hence, for a given sampling frequency F_{samp}, the minimum distance between two detectable nodes is equal to

$$d = \frac{0.3\, C}{F_{samp}}. \tag{3.6}$$

This path delay detection resolution can be used to assess the capability of a given PLC system in detecting adjacent nodes. Actually, the existing PLC standards are deploying carrier frequencies from a few kHz on the Narrowband Power Line Communication (NB-PLC) to MHz range in the BB-PLC [250]. It is evident that Broadband Power Line Communication (BB-PLC) offers, by default, a better detection accuracy than that of NB-PLC. As an example, using a sampling frequency larger than 200 MHz [211, 254] results in a detection resolution below 50 cm.

3.5 Additive Noise in PLC

In PLC, there are two main sources of noise: (1) adjacent electrical activities and (2) electromagnetic radiation absorbed by the power lines themselves. These generated noise can be classified into three categories: (i) base-band impulsive noise that is generated by nonlinear circuit elements consuming electrical current [155]; (ii) continuous impulsive noise that is harmonically related to the fundamental frequency, which is present during the operation of electromechanical devices, e.g., motors and silicon-controlled rectifiers; and (iii) transient impulsive noise that is typically observed during the switching activities of power cycle devices. This type of noise generally lasts for a brief span of milliseconds [51].

Furthermore, a wide range of electromagnetic emissions, including radio transmission, can affect the PLC, where power lines act as efficient receiver antennas in this scenario [253].

Chapter 4
Power Line Communication Security

Abstract This chapter provides an in-depth exploration of the fundamental principles and vulnerabilities associated with Power Line Communication (PLC) security. It also explains the key security attributes of PLC networks, including confidentiality, integrity, availability, and authentication, along with the associated challenges.

4.1 PLC Issues and Limitations

Various challenges pervade the PLC landscape, each potentially significantly impacting its performance and integrity across multiple dimensions. The following section categorizes these challenges into three main areas: technical obstacles, security vulnerabilities, and network-related issues. Each category presents its own intricacies that require tailored solutions for optimized and robust PLC performance.

- **Technical issues**
  - **Behavioral issues:** A change in PLC device, application, or system performance may manifest due to irregular activities. These activities can range from malicious cyberattacks, such as unauthorized data access or system interference, to unidentified technical complications, such as software glitches or hardware malfunctions. These deviations may compromise the system's intended functionality and could necessitate immediate investigation and remediation to restore normal operation.
  - **Design issues:** Balancing security and performance in PLC design is critical. Often, one aspect is prioritized at the expense of the other, leading to compromises that may not be suitable for real-world applications. Achieving a proper balance between security measures and performance capabilities is crucial to protecting users' safety and privacy.
  - **Compatibility issues:** These arise when new or legacy technological standards and protocols are implemented without adequate evaluation. Such a mismatch can impose limitations on the PLC system, including performance degradation and regulatory compliance issues.

J. Hernandez Fernandez et al., *Physical Layer Security in Power Line Communications*, Advances in Information Security 108,
https://doi.org/10.1007/978-3-031-57349-1_4

  - **Accidental issues:** These usually stem from human errors, such as inadequate training or mistakes made by personnel. If not promptly addressed, such issues can escalate into more serious problems, potentially causing cascading failures across the system.
  - **Voltage issues:** These are associated with irregularities in the incoming or outgoing current/voltage levels, affecting the overall performance of the PLC system. Voltage issues impact the PLC system and may jeopardize connected infrastructure, such as smart cities and buildings, by damaging electronic and electrical devices.

- **Security issues** Security issues encompass a range of vulnerabilities targeting diverse security breaches. One illustrative instance is the impact of low line impedance on the analog front-end design of a PLC modem. When located at the transmitter port, this creates difficulties for voltage signal injection [62].

  - **Signal interception:** This may arise due to unauthorized surveillance techniques like eavesdropping, wiretapping, or packet sniffing, particularly when the PLC system transmits unencrypted data over insecure communication channels. Such activities compromise both the system's confidentiality and the privacy of its users.
  - **Signal interruption:** This primarily occurs through various forms of cyber-attacks, such as man-in-the-middle, replay, wormhole, sinkhole, blackhole, Sybil, or DoS/DDoS attacks. These attacks can interfere with, modify, or delay the transmission or reception of PLC packets. This vulnerability often arises due to inadequate Intrusion Detection Systems (IDS), firewalls, and secure communication channels (e.g., frequency hopping, multiplexing).
  - **Deployment and performance issues:** The integration of certain security measures, whether cryptographic or non-cryptographic, can adversely affect the performance of resource-limited PLC systems and devices. The strategic placement of such security solutions can also present challenges that compromise overall system efficiency. Consequently, a balanced approach necessitates lightweight yet effective security solutions.
  - **Safety issues:** It is crucial to differentiate between safety and security; the former pertains to minimizing hazards that can lead to physical harm or material loss, while the latter addresses protection against unauthorized access and data breaches. PLC systems must, therefore, be designed to minimize risks to personnel and provide emergency shutdown capabilities.
  - **Signal injection issues:** Malicious software can intercept and alter the PLC payload by injecting malicious code. This tampering compromises various aspects of PLC performance, such as availability and decision-making. To counter this, appropriate security mechanisms like intrusion detection systems, honeypots, and artificial intelligence-based solutions are essential.
  - **Authentication issues:** If not adequately managed, authentication protocols can be exploited by insiders or rogue employees attempting to gain unauthorized access, damage PLC devices, or intentionally disrupt system operations. Multifactor authentication strategies, including device tokens,

biometric verification, and access privilege management, are imperative for mitigating such risks.

- **Network issues** Network exploitation is contingent on various factors, such as the network type, characteristics, and established standards within the PLC environment [56, 166]. In this context, Table 4.1 outlines a range of network vulnerabilities that could potentially compromise PLC systems.

4.2 PLC Security Threats

Wired and wireless channels in a smart grid environment are susceptible to a range of threats and vulnerabilities, both cyber and physical, that could compromise them [215]. The impact of such attacks on PLC systems may vary; however, the principal aims of these assaults typically concentrate on compromising confidentiality, integrity, availability, and authentication, as elaborated below. Attacks on the electrical grid originate from infrastructural shortcomings, either deliberate or accidental, leading to cybersecurity lapses that can trigger a range of adverse outcomes, including cascading failures, blackouts, sags, brownouts, spikes, surges, inrush currents, noise, and transient issues [70]. Furthermore, PLC systems are not limited to grid applications; they are also employed in maritime, aerospace, and automotive settings, adding additional layers of vulnerability and complexity [62, 251]. The potential areas of attack can include:

- **Confidentiality:** Multiple vulnerabilities jeopardize the confidentiality of PLC. These systems are susceptible to various forms of intrusive activities such as war-dialing, wiretapping, and piggyback attacks that compromise data and information during transmission. This opens up opportunities for sabotage or espionage. Additionally, PLCs are at risk from port scans, reconnaissance, and eavesdropping attacks, which intercept real-time data and communication streams. The ongoing challenge is to implement effective solutions that safeguard the confidentiality of PLC systems in a real-time, secure environment. Utilizing outdated or weak cryptography algorithms, like the Data Encryption Standard (DES), exposes PLC systems to an array of cryptographic and statistical attacks. Privacy intrusions can lead to erroneous decision-making and actions based on faulty data, real-time surveillance, and the disclosure of user activities [212]. Such breaches could also facilitate electricity theft by manipulating meter readings or modifying information via compromised encryption keys [67].
- **Integrity:** Threats to logical integrity can culminate in sabotage attacks, potentially causing physical harm or even the destruction of PLC systems. Additionally, man-in-the-middle attacks can tamper with data as it is transmitted, altering its content. Consequently, a pressing issue is to identify an effective hashing solution to fortify the integrity of PLC systems against these and similar vulnerabilities.

Table 4.1 A set of exploitable network vulnerabilities [250]

Network vulnerabilities	Description
Destination Address Alteration (DAA)	DAA attacks occur when the destination address field is changed by the attacker, which allows him to reroute the request(s)/reply(ies) to other devices, resulting into unexpected suspicious results
Length Overflow (LO) and Dynamic Flow Concept (DFC)	LO & DFC attacks can result in inserting wrong or fake length filed value. This can seriously affect either the message processing or the DFC flag. As a result, the affected devices will appear busy to the master network while their data is being corrupted and manipulated, causing the devices to crash unexpectedly
Passive Network Reconnaissance (PNR)	PNR attacks occur when an attacker is capable of capturing and analyzing vendor-specific protocol messages. Such attacks can sniff and leak sensitive/crucial information about the topology of a network, the device's functionality, and other data
Transport Sequence Modification (TSM)	TSM attacks occur when malicious data is injected through the insertion of fabricated messages into the sequence of a fragment
Rogue Commands (RC)	RC includes messages that hackers send to outstation devices to affect their devices negatively
Rogue Interloper (RI)	RI attacks occur when a third-party device is installed between the bus station and the master to launch a man-in-the-middle attack that allows the attacker to read, modify, and fabricate messages and/or network traffic
Radio Frequency (RF) and Unavailable Function (UF)	RF & UF attacks aim to force the devices to restart. This will affect the system availability since it will take the devices an additional time to restart. Devices might also be interrupted, disrupted, and modified due to a possible malicious data injection

- **Availability:** Vulnerabilities may arise from malicious actions such as physical tampering with PLC cables (e.g., severing or damaging them) or through denial-of-service assaults. Natural events, like extreme voltage fluctuations, can also critically hinder the regular functioning of PLC systems. Furthermore, power outages can inflict substantial harm, particularly when secondary backup power solutions are lacking. Consequently, a key issue is the appraisal and implementation of robust secondary backup hardware and cabling systems to ensure sustained availability.
- **Authentication:** Risks originate from two primary elements: the source, which may be untrustworthy and malevolent with the aim of inflicting damage, and the user, who may misuse or escalate privileges to initiate an insider assault or divulge confidential information. Accordingly, upholding robust mechanisms for both source and user authentication poses a significant challenge, particularly in the current climate where acts of espionage and sabotage are increasingly prevalent.

In Fig. 4.1, we outline the principal Narrowband Power Line Communication (NB-PLC) protocols, specifically, G.hnem [30, 120, 202], G3-PLC [205], and PRIME [75]—as well as Broadband Power Line Communication (BB-PLC) protocols such as G.hn [27], G.Hn MIMO [230], HD-PLC [64, 225], and Home Plug AV/GP/AV2 [54, 225].

4.2.1 Threats' Sources

Numerous techniques can be employed to compromise a specific PLC system and its devices, as outlined in the following points [71]:

- **Malicious agents**, mainly internal actors, harbor malicious intentions and objectives, detailed further in [71, 180]. Their motives can span diverse areas such as cybercrime [180], cyberterrorism [135], cyber warfare [244], cyber-politics associated with hacktivism [15, 119], and industrial espionage [31].
- **Botnet** herders may take advantage of susceptible PLC devices by initially deploying spyware or malware for exploitation, before orchestrating a coordinated remote assault aimed at compromising the availability of the PLC systems.
- **Spies** often have the objective of infiltrating PLC systems, initially conducting clandestine reconnaissance to assess the vulnerabilities of a specific PLC setup and/or system, with the intent to exploit these vulnerabilities at a subsequent stage [47]. Such spies may be involved in cyber-espionage or industrial/economic espionage activities [136, 228] and are frequently associated with broader global cyber-political conflicts [243]—such as Titan Rain (2003), Operation Olympic Games (2006), GhostNet (2009), and Shadow Network (2009) [125, 167]).
- **Cyber-military operations** operations often prioritize the targeting of civilian infrastructure—such as ports, telecommunication systems, power plants, bridges, and occasionally hospitals [248] as a prelude to full-scale military activities,

Fig. 4.1 Existing security solutions for PLC standards [250]

including invasions or comprehensive conflicts [69]. Notable examples include Iraq in 2003 [36, 124], Lebanon in 2006 [126], and Gaza in 2008 and 2014 [24, 242]. This strategy has evolved to encompass modern forms of cyber-warfare [152, 213], where hacking serves as a cyber-weapon aimed at various cyber-physical infrastructures [249]. Noteworthy instances of such cyberattacks include Estonia in 2007 [179], Georgia in 2008 [53], Ukrainian power distribution systems in 2015 via the BlackEnergy malware [137, 224], and Iran in 2010 and 2020 [25, 45, 139].
- **Criminal groups** engage in unauthorized access to specific PLC devices, substations [95], or entire power grids. Their actions, which may include the disruption or jamming of electrical power transmissions, are often motivated by hacktivism or personal financial benefit.
- **Insiders** are individuals within an organization who have access to sensitive information or systems and are often motivated by factors like lust, greed, or dissatisfaction. These individuals can participate in sabotage activities that negatively affect the PLC environment.
- **Phishers and spammers** initiate their attacks with a reconnaissance phase, during which they collect publicly accessible information about a targeted PLC system. Subsequently, they disseminate malicious emails containing either harmful links or attachments. Once activated, these elements may compromise PLC devices and collect sensitive data such as passwords, layout maps, and details. This information is highly valuable to the attacker for purposes of sabotage, as evidenced in incidents like Night Dragon (2009), Shamoon (2012), Greek Oil Campaign (2012), and New Year's Campaign (2013) [82, 246].
- **Terrorists** focus on attacking critical infrastructure components, such as power stations [69], employing methods ranging from cyber-hacking to kinetic assaults like bombings or the use of armed drones [247]. Illustrative of this is the 2019 Abqaiq–Khurais attack [97]. The primary objective of these actions is to inflict maximum disorder, devastation, and disruption [180, 216].

4.2.2 Threat Types

Various forms of threats exist that can be specifically classified in the context of PLC systems as follows:

4.2.2.1 Malware

Categorized as malicious software, these programs manifest in diverse forms, such as adware, spyware, ransomware, or Trojan horses. Each serves a unique function and poses specific threats to system security, ranging from unauthorized data collection to system damage. The following sections provide an in-depth examination of these various types:

- **Trojans** are malevolent software elements that infiltrate PLC systems under the guise of legitimate applications, such as those for monitoring PLC circuits or assessing system performance. These Trojans can compromise hardware by manipulating the electrical circuitry within the PLC, resulting in payload transmissions that exceed or fall below anticipated levels [122, 168].
- **Viruses** are a type of malicious code that, when executed, replicates by modifying other computer programs and inserting its own code. In contrast to worms, they require human intervention to initiate their clandestine proliferation. They target vulnerable PLC devices and systems intending to compromise their availability, integrity, security, safety, and performance [13].
- **Worms** are sophisticated forms of viruses capable of autonomous propagation, thereby enabling more rapid dissemination. This category of malware is commonly employed to compromise industrial sectors, including PLC systems, as exemplified by instances like Stuxnet 1.0 and 2.0 [23, 147].
- **Logic bombs** consist of code segments embedded within a vulnerable PLC software system. These are programmed to activate at a predetermined time or under specific conditions, intending to disrupt the system's functionality by deleting, intercepting, or altering data and processes.
- **Spyware** serves as a surveillance mechanism, designed to capture and intercept PLC transmissions for subsequent scrutiny and potential modification.
- **Ransomware** has the capability to immobilize PLC devices and systems by encrypting their files through malware dissemination, thereby obstructing regular access and functionality.

4.2.2.2 Malicious Applications

The application layer of the PLC is susceptible to a range of preexisting attacks targeting either its firmware or software components. Such attacks encompass the following:

- **Misconfiguration** involves issues such as erroneous code, outdated patches, and weak coding practices. These vulnerabilities can lead to cascading failures in PLC devices, potentially culminating in a complete blackout or system outage.
- **Web interface Web Interface** encompasses various types of attacks, Structured Query Language Injection (SQLi), IP misconfiguration, Cross-Site Scripting (CSS), and Cross-Site Reference Forgery (CSRF) [231]. These vulnerabilities can jeopardize the integrity, privacy, and authenticity of PLC systems by illicitly gaining access to confidential and sensitive data stored in the devices, such as electrical substations or connected infrastructures. Such breaches may form part of cybercriminal or cyber-espionage activities [39, 55].
- **Buffer overflow attacks** aim to compromise the availability of PLC systems. Typically, these attacks manipulate the internal network tables of the PLC, altering the execution pathways of its programs. Such manipulations can induce

aberrant or malicious responses, thereby endangering files or revealing confidential information [76, 123].
- **Zero-day exploit** involves taking advantage of undisclosed vulnerabilities and security lapses within PLC systems to target specific devices, such as electrical substations, smart meters, or electric vehicle chargers. The primary objectives are to elevate the attacker's permissions to administrative levels or to disrupt standard operational protocols. These exploits are commonly employed in cyber-espionage activities, exemplified by instances like Stuxnet 1.0 and 2.0, Flame, Duqu 1.0 and 2.0, and Gauss [20, 120].

4.3 PLC Security Attacks

PLC networks are often structured utilizing a Network Encryption Key (NEK). Given that power line channels are shared networking mediums, they are susceptible to external and internal security threats and attacks [26]. Recent increases in attacks targeting Network Membership Key (NMK) within PLC systems have been noted [146]. These incursions can be attributed to either external or internal actors or a combination of both [44]. The primary forms of these attacks are as follows:

- **Interception** These comprise passive eavesdropping to listen to broadcast packets, as well as active interference such as man-in-the-middle attacks executed through counterfeit access points or manipulated MAC/IP addresses.
- **Sniffing** This entails either passive or active interception of encryption key packets with the aim of deciphering the key combination and capturing transmitted data. Moreover, the copper wiring of PLC systems can be subjected to wiretapping as a means to access or falsify the data being transmitted [49].
- **LAN attacks** This situation arises when a device is locally set up without the utilization of any Direct Access Key (DAK), connected to a switch or router. Such a configuration leaves the system vulnerable to various types of network attacks, including backdoor incursions [54], Domain Name System (DNS) spoofing [237], port scanning [132], and denial of service assaults [131], among others [48].
- **Brute-force/dictionary** This approach involves an exhaustive password-cracking strategy aimed at compromising the authenticity of PLC systems. Specifically, brute-force attacks target the NMK with the intent of obtaining the key to decrypt both transmitted and received packets, undermining the confidentiality, integrity, and privacy of the conveyed information [54]. Countermeasures for such attacks may include the implementation of a robust user password policy and multifactor authentication.
- **Identity forgery** This takes place when adversaries position themselves intermediate to the HAN and the NAN, effectively serving as a conduit for intercepted communications [26].

- **Data attacks** Operations within the power market are vulnerable to data integrity attacks. Here, assailants seek financial gains from market transactions via the injection of fraudulent data [141]. This enables them to subvert multiple sensor lines without raising alarms [84].
- **Load alteration attacks** are designed to impact the consumption sector by adjusting electrical loads at predetermined grid locations. The goal is to compromise the electrical grids by causing circuit overflows, thereby unsettling the power supply and demand balance. These attacks can be subdivided into two main categories:

  1. **Direct load control**: Load management systems feature bidirectional communication, allowing for the transmission of command signals either through Internet channels [143] or via power line carriers [241]. This dual-path communication is susceptible to unauthorized manipulation, where attackers can compromise the command signals to adjust vulnerable electrical loads, thereby causing a noticeable increase in total demand. More advanced versions of this threat include temporal noise manipulation, known in the literature as dynamic load fluctuation attacks [11].
  2. **Indirect load control**: This is an alternative strategy to direct load control within demand-response frameworks. In this approach, load adjustments are mediated by utility companies using real-time pricing models, often communicated over the Internet. Automation of indirect load control can be accomplished through smart meter-based energy consumption schedules or by timing the energy use of individual controllable household appliances [162].

Chapter 5
Physical Layer Security in Power Line Communications

Abstract The field of Physical Layer Security (PLS) applied to Power Line Communication (PLC) networks remains largely untapped. This section presents an overview of the foundational concepts of PLS and its distinct implementations within PLC. The discussion then shifts to various PLS techniques found in the academic literature. Each technique's applications, benefits, and constraints are classified, accompanied by relevant performance analysis metrics for a thorough evaluation.

5.1 Physical Layer Security

PLS has emerged as a transformative approach in the field of network security, uniquely exploiting the physical attributes of the communication medium to facilitate secure data transmissions. Unlike traditional cryptographic methods that operate at the application, transport, or network layers, PLS utilizes the physical layer's inherent features for security provisions. In various communication environments, such as wireless networks, optical communications, and PLC, the physical layer presents inherent randomness and environmental noise that can be harnessed for security purposes. Specifically, the characteristics of the channel, like fading, interference, signal attenuation, and spatial configurations, can be employed to generate encryption keys or thwart unauthorized access [10, 142].

The advantage of PLS is its ability to offer an added layer of security that operates independently from classical encryption protocols. This makes it particularly effective in scenarios where computational limitations or protocol constraints may impede the application of traditional cryptographic methods. For instance, in low-power Internet of Things (IoT) devices or real-time communications, the processing overhead of implementing complex encryption algorithms may be prohibitive [165]. Here, PLS can provide an efficient and effective security layer.

The inherent robustness of PLS against computational attacks is a cornerstone of its effectiveness. For instance, when it comes to key generation, this robustness is attributable to the fact that security keys are generated based on the unique temporal

J. Hernandez Fernandez et al., *Physical Layer Security in Power Line Communications*, Advances in Information Security 108,
https://doi.org/10.1007/978-3-031-57349-1_5

or spatial characteristics of the physical layer, such as channel fading, interference patterns, or multipath propagation effects. These nuances are inherently difficult, if not impossible, for an eavesdropper to replicate without physical proximity and precise synchronization with legitimate users [133]. Even if an attacker was able to approximate these features, the intrinsic randomness and transient nature of the physical layer make it exceptionally challenging to capture these attributes accurately enough to compromise the system.

In traditional cryptographic systems that operate at higher layers, such as the application or transport layer, encryption keys are often based on algorithmic computations that, in theory, could be reverse-engineered or cracked through brute-force attacks given enough computational power and time [250]. PLS, on the other hand, sidesteps this vulnerability by rooting its security in the physics of the communication medium itself. Here, the transience and variability of physical layer attributes act as an ever-changing encryption algorithm highly resistant to computational attacks.

The integration of machine learning and statistical methods with PLS represents a significant advancement in the field, allowing for a more dynamic and adaptive security posture. Traditional PLS methods often rely on predefined algorithms and heuristics that, although effective, are static in nature. They might not readily adapt to rapidly changing channel conditions or emerging security threats. In contrast, machine learning algorithms can be trained to recognize and adapt to nuanced changes in channel behavior, environmental variables, and other contextual factors that could affect the security of a communication system.

For example, machine learning models such as neural networks, decision trees, or support vector machines can be employed to continuously analyze the historical and real-time data from the physical layer. These models can identify subtle patterns and correlations that may not be immediately obvious but are crucial for the effective generation and maintenance of secure communication links [88]. This capability is particularly important in scenarios where the physical layer is subject to frequent changes due to mobility, interference, or varying transmission conditions [89].

Additionally, statistical methods can offer a deeper understanding of the channel's behavior by quantifying its various attributes, such as variance, skewness, or kurtosis. These insights can then be used to fine-tune the parameters of the PLS system, optimizing it for the prevailing channel conditions. For instance, if statistical analysis reveals a greater level of noise or interference at certain times, the PLS system can adjust its key generation algorithms to maintain a high level of security, even under suboptimal conditions.

The real-time adaptability enabled by machine learning and statistical methods grants a system the ability to adjust its security protocols on the fly. Whether responding to varying channel conditions or adapting to new types of security threats, this level of agility is a strong complement to traditional cryptographic systems, which may not be as flexible or quick to adapt.

This allows for an agile security posture that can adjust to varying channel conditions or emerging threats, further strengthening its role as a complementary security measure to traditional cryptographic systems.

5.2 Physical Layer Security in PLC

While PLC systems have many benefits, such as cost-efficiency and easy deployment, they also come with certain security concerns. The inherent broadcasting nature of PLC networks, coupled with their extensive reach via the electrical grid, increases their susceptibility to unauthorized access and malicious actions [103]. These factors create a favorable environment for eavesdroppers and malicious actors to exploit the vulnerabilities of PLC systems, potentially leading to a wide array of detrimental consequences such as data theft, unauthorized control over smart grid functions, or even infrastructure sabotage [250].

Moreover, these vulnerabilities become even more critical when PLC systems are deployed in sensitive applications such as industrial control systems, healthcare monitoring, and emergency services, where a security breach could have catastrophic implications. For instance, interference with PLC communications in a healthcare setting could jeopardize patient safety by causing malfunctions in medical devices or disrupting essential services.

To address these challenges and enhance the security of PLC systems, it is imperative to implement robust security measures tailored to their unique characteristics. While conventional cryptographic methods provide a foundational level of protection, they may encounter limitations due to the distinctive properties of the PLC channel. Cryptography has long served as the bedrock for safeguarding data communication [40, 170, 236]. As previously elaborated and depicted in Fig. 4.1, PLC standards exemplify this trend by mainly utilizing Public Key Infrastructure (PKI) mechanisms for their security protocols.

Traditionally, pre-shared keys have been employed to ensure the confidentiality of transmitted data, making decryption practically infeasible without access to these keys [229]. However, certain cryptographic algorithms, especially those reliant on public key infrastructure, demand substantial computational resources [170]. Furthermore, as the computational capabilities available to potential eavesdroppers continue to advance and with the looming prospect of quantum computing, conventional encryption techniques are becoming progressively vulnerable to security breaches [127, 229].

To enhance the security framework of PLC systems, PLS serves as an effective tandem with traditional encryption methods [32]. The core concept of PLS leverages unique features of the communication channel, such as fading, noise, or Signal-to-Interference-Plus-Noise Ratio (SINR), to facilitate secure and efficient communication between a transmitter and the intended receiver, irrespective of the computational capacity of potential eavesdroppers [87, 196].

The following part of this section will offer a synopsis of the studies in the existing research body that have investigated diverse elements of PLS in the context of PLC. This will be followed by a subsequent section dedicated to summarizing the PLS techniques in PLC presented in the academic literature.

In the research article cited as [33], the authors carried out field experiments in several Brazilian residences to evaluate the eavesdropping resilience of PLS functionalities within a PLC system. The study aimed to evaluate the effectiveness of secure communication between an authorized PLC transmitter and a legitimate PLC receiver. Various metrics, such as the Secrecy Rate (SR), the Secrecy Outage Probability (SOP), the Effective Secrecy Throughput (EST), and the Wiretap Code Rate (WCR), were utilized to quantify the PLC system's security performance across three distinct frequency bands: 1.7–30 MHz, 1.7–50 MHz, and 1.7–86 MHz.

In a comparative analysis, the research outlined in [191] assesses the Average Secrecy Capacity (ASC) in both conventional wireless and PLC channels subject to fading. The findings corroborate that wireless channels generally yield a higher ASC compared to PLC channels. This discrepancy can be attributed to variations in channel statistics and propagation environments. To validate these theoretical insights, empirical measurements within residential PLC setups were carried out and examined.

Continuing in a similar vein, the authors in [192] extended their expertise in wireless fading channels to standard PLC settings. Initially, they conducted an in-depth examination of PLC channel statistics, contrasting them with those of wireless channels. Subsequently, leveraging both empirical in-home PLC data and channel simulations, the secrecy rate of a specified tree topology within the PLC network was assessed.

In a subsequent study [194], the authors explored transmission security performance within in-home networks using measured noise and channel responses. Their investigation focused on the achievable secrecy rate and took into account various factors such as channel properties, transmission bandwidth, noise type, and transmit power. These findings were then juxtaposed with the conventional wireless scenario, where channels are commonly assumed to be independent and Rayleigh-distributed. Moreover, the potential for performance improvement through the utilization of Multiple-Input and Multiple-Output (MIMO) transmission was also examined. Consequently, factors such as impulsive noise, user proximity, path loss coefficients, and transmission power have been explored through both theoretical analysis and Monte Carlo simulation results.

The effects of channel reciprocity on the performance of PLS in Time-Division Duplexing (TDD) and Frequency-Division Duplexing (FDD) systems have been assessed in [85]. The research concentrated on the shared randomness that can be employed in the generation of physical layer keys.

In an alternate setting, the study in [32] examines the security performance of a PLC system when confronted with a rogue wireless communication device. This security vulnerability manifests when unshielded power cables are utilized in PLC systems. Utilizing a dataset of channel estimations and additive noise from a residential measurement campaign, the study offers empirical analysis and valuable insights. The findings indicate that the Security Outage Probability (SOP) is considerably elevated when the PLC devices are separated by approximately 6 meters, and the eavesdropper is within 2 meters of the PLC transmitter. As a result, reevaluating the use of unshielded power cables or exploring new design paradigms for broadband in-home PLC systems becomes crucial for ensuring secure PLC operations. As a result, rethinking the use of unshielded power cables or innovating new architectural designs for broadband in-home PLC systems is essential for enhancing PLC security.

5.3 Physical Layer Security Techniques in PLC

In this section, we detail the presented PLS techniques discussed in the academic literature, organizing them according to their processing domain and the performance analysis metrics employed for their evaluation. A consolidated summary of these techniques is presented in Table 5.1.

5.3.1 Time Domain

In [186], the authors explore the notion of secure physical layer key generation within the context of asymmetric PLC channels. Specifically, they introduce two PLS key generation methodologies utilizing distinct quantization techniques. To assess the security efficacy of these approaches, the spatial correlation between the primary and wiretap channels is investigated.

In a related study, the spatial correlation within a specified adversarial PLC system model is thoroughly examined in [86]. Initial findings indicate that PLC channels possess slow time-varying characteristics. As a response, the authors introduce an innovative method aimed at enhancing the randomness of PLS keys, thereby reducing the spatial correlations between the main and wiretap channels. This strategy involves the use of idle branching connections equipped with random (reactive) loads. The findings demonstrate a notable difference between the main and wiretap channels, even when the eavesdropper is situated merely a meter away from the legitimate transmitter.

In a separate study [87], the authors devised a symmetric virtual Channel Impulse Response (CIR) to counter the inherent asymmetry in PLC Channel Frequency Response (CFR)s, an issue that has constrained the progress of physical layer techniques. To this end, they introduced a low-complexity, multilevel quantization algorithm for PLS key generation, which achieves a Bit Mismatch Rate (BMR). The performance of this innovative approach was evaluated by comparing its BMR to that of single-level quantization across various topologies. The findings suggest that the new technique holds promise for effective PLS implementation.

In a related study [259], the authors introduced a novel technique for PLS key generation that addresses the limitations of PLC channels, specifically their slow time-varying characteristics and additive noise. The proposed methodology employs a CFR quantization algorithm, which adapts both amplitude threshold and quantization level relative to the noise power, aiming to optimize the Key Generation Rate (KGR). The efficacy of this approach was scrutinized under two different adversary scenarios: a single eavesdropper and multiple eavesdroppers. The findings

Table 5.1 PLS techniques for PLC according to their processing domain and the performance analysis metrics employed for their evaluation [178]

					PLS performance analysis metrics										
Domain	Techniques	Advantages and disadvantages	Ref	Apps	SpCr	ASC SR	EST ST	SOP	SPDP	BAR BMR	SEE	BGR	KGR KDR	CDR	RND
Time	PLS key generation	+ A powerful security technique. – Increasing the transmission latency. – Sensitive to the CSI estimation error.	[186]	Prevent eavesdropping attacks	✓										
			[86]	Prevent eavesdropping attacks	✓										
			[87]	Prevent eavesdropping attacks						✓					
			[259]	Prevent eavesdropping attacks									✓		
			[253]	Prevent eavesdropping attacks						✓		✓	✓		
			[89]	Topology change detection										✓	
	PHY-Au based on PLC noise	+ Efficient-cost IoT device authentication. – Limited to local power line only.	[133]	Pairing and authentication						✓					
	PHY-Au based on CSI	+ Accurate identification. – Multiple PLC nodes are needed.	[88]	Authentication					✓						

Frequency	Friendly jamming	+ Increasing the secrecy energy enhancement. –A high transmission latency.	[196]	Prevent eavesdropping attacks		✓					✓				
	PHY keyed-hash function	+ Robust and efficient technique. –A high transmission latency.	[153]	Source authentication and message integrity											✓
	PLS key-based IDA	+ A low overhead and a high degree of availability and reinforcing confidentiality. –A high transmission latency.	[169]	Data availability and confidentiality											✓
	DCSK	+ A powerful security technique. – A low spectrum efficiency.	[160]	Prevent eavesdropping attacks		✓	✓	✓							
			[161]	Prevent eavesdropping attacks			✓	✓							
	CFR-based PLS key	+ A powerful security technique. – Sensitive to the CSI estimation error.	[186]	Prevent eavesdropping attacks	✓										

(continued)

Table 5.1 (continued)

					PLS performance analysis metrics										
Domain	Techniques	Advantages and disadvantages	Ref	Apps	SpCr	ASC SR	EST ST	SOP	SPDP	BAR BMR	SEE	BGR	KGR KDR	CDR	RND
Spatial	Cooperative PLC	+ A powerful security technique. – A low spectrum efficiency.	[209]	Prevent eavesdropping attacks		✓									
			[210]	Prevent eavesdropping attacks		✓									
	Power allocation	+ A high secrecy energy efficiency. – A high transmission latency.	[196]	Prevent eavesdropping attacks		✓					✓				
			[255]	Prevent eavesdropping attacks		✓									
			[3]	Prevent eavesdropping attacks				✓							

SpCr spatial correlation, *ASC* average secrecy capacity, *SR* secrecy rate, *ST* secrecy throughput, *EST* effective secrecy throughput, *SOP* secrecy outage probability, *SPDP* successful path detection probability, *BAR* bit agreement rate, *BMR* bit mismatch rate, *SEE* secrecy energy efficiency, *BGR* bit generation rate, *KGR* key generation rate, *KDR* key disagreement rate, *CDR* connection detection rate, *IDA* information dispersal algorithm, *DCSK* differential chaos shift keying, *RND* randomness affirm the robust security performance of the proposed framework in terms of both KGR and Key Disagreement Rate (KDR).

In an effort to enhance the security of PLC IoT devices, the study by [253] introduced a technique that leverages the shared Electromagnetic Interference (EMI) spike frequencies among outlets within a verified electrical domain. The effectiveness of this methodology was assessed using various performance indicators, such as BMR, BGR, and KDR.

In the study [88], a novel physical layer identification mechanism for PLC systems is elaborated, which operates under imperfect Channel State Information (CSI) conditions. This methodology employs a multipath channel delay detection technique to achieve precise physical layer identification for the PLC links in question. The performance of this approach is validated through the evaluation of successful Successful Path Detection Probability (SPDP) under both ideal and compromised CSI conditions.

In [89], a pioneering application for power grid monitoring is presented, leveraging CIR-based keys to identify alterations in grid topology. The efficacy and practicability of the proposed methodology were demonstrated through a real-world implementation and a comprehensive series of simulations. The findings validate the system's robustness by confirming a high Connection Detection Rate (CDR), even in the presence of significant noise interference.

The concept of physical pairing and authentication is another PLS technique introduced in [133] aimed at autonomously securing device-to-device interactions among IoT devices via PLC noise-based key generation. The central premise revolves around harvesting the shared PLC noise randomness between proximate devices to safely generate and utilize a PLS key. Subsequent experiments across diverse settings were conducted to substantiate the efficacy of the approach in excluding unauthorized devices lacking access to the primary local power line.

5.3.2 Frequency Domain

The concept of enhancing the security of MIMO PLC systems through the use of full-duplex jamming receivers is presented in [196]. Employing In-Band Full-Duplex (IBFD) technology, the proposed approach enables simultaneous reception of the intended message and jamming of the operational frequency band. The authors have fine-tuned both the information and jamming transmission strategies, along with power allocation across various subcarriers, to optimize Secrecy Energy Efficiency (SEE). Computational results were provided to corroborate the efficacy of the suggested full-duplex strategy in contrast to a half-duplex operation lacking jamming, particularly in terms of SR and SEE.

In [153], the authors unveil a novel hashing methodology applied to OFDM frame symbols, leveraging the random attributes inherent in the physical layer. A comprehensive security and performance analysis substantiates that the proposed system ensures elevated levels of both message integrity and source authentication.

In a subsequent work presented in [169], the authors advocate for the employment of the Information Dispersal Algorithm (IDA) in conjunction with the characteristics of PLC channels. This combination aims to achieve heightened availability while fortifying confidentiality, all with minimal overhead and computational complexity.

In [160], the authors explore the effectiveness of using the Differential Chaos Shift Keying (DCSK) modulation technique to safeguard authorized users from multiple eavesdropping attempts. The study quantitatively assesses the efficacy of this approach by focusing on key metrics such as the ASC, the SOP, and the ST. An extension of this work is presented in [161], where the authors evaluate the PLS effectiveness of different DCSK modulation schemes within PLC frameworks. The study employs the SOP and secrecy throughput metrics as performance indicators to validate the security robustness of these implementations.

5.3.3 Spatial Domain

In [209, 210], the authors evaluate the PLS of cooperative relaying PLC systems when an eavesdropper is present. Performance is assessed using the ASC as the key metric. Similarly, [255] discusses an optimal power allocation strategy aimed at maximizing the SR within a broadband, two-hop multicarrier relay PLC system. This notion of power allocation is further explored in [3], where the paper addresses secrecy outage analysis in Non-Orthogonal Multiple Access (NOMA) PLC systems. The findings indicate that both impulsive noise and the adopted power allocation methodology significantly influence the reduction of the SOP.

Chapter 6
Performance Analysis of Data Transmission Security and Reliability in PLC

Abstract This chapter investigates well-known Physical Layer Security (PLS) performance analysis metrics that can be used to evaluate a variety of PLS techniques for Power Line Communication (PLC) networks. Furthermore, a new PLC performance analysis metric, named Secure and Successful Transmission Probability (SSTP), has been introduced and explained. This metric aims to assess both the security and reliability of a given transmission link. Moreover, for each metric, the main concept and some performance analysis numerical results in real PLC scenarios are presented and evaluated.

6.1 System and Adversary Model

This section details the system and adversary model under consideration. As presented in Fig. 6.1, the system model consists of a transmitter $Alice$ (A) that sends private messages to a legitimate receiver Bob (B), in the presence of a malicious Power Line Communication (PLC) device Eve (E) that is connected to the same PLC network. In this scenario, Eve may carry out a variety of security attacks, including jamming, spoofing, and eavesdropping.

Throughout this chapter and without loss of generality, we focus on an eavesdropping attack scenario, where Eve intends to spy upon private communications between $Alice$ and Bob.

In this case, the corresponding PLC wiretap channel model is presented in Fig. 6.2, where h_{A-B} and h_{A-E} are the complex channel gains of $Alice$-Bob link and $Alice$-Eve link, respectively. And n_{A-B} and n_{A-E} are the relevant additive noise. Accordingly, the general expression of the instantaneous Signal-to-Noise Ratio (SNR), denoted by γ_{A-X}, $X \in \{B, E\}$, is given by

$$\gamma_{A-X} = \frac{P_T \, H_{A-X}}{N_X}, \tag{6.1}$$

J. Hernandez Fernandez et al., *Physical Layer Security in Power Line Communications*, Advances in Information Security 108,
https://doi.org/10.1007/978-3-031-57349-1_6

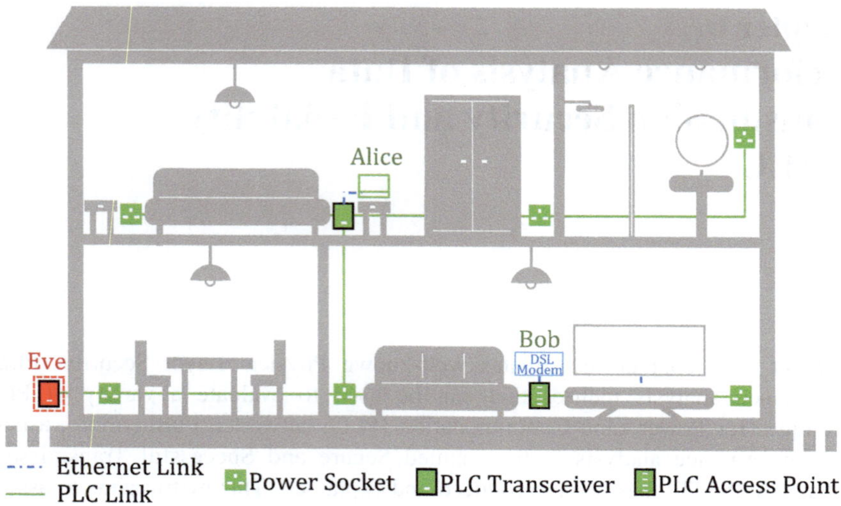

Fig. 6.1 The adversary system model [178]

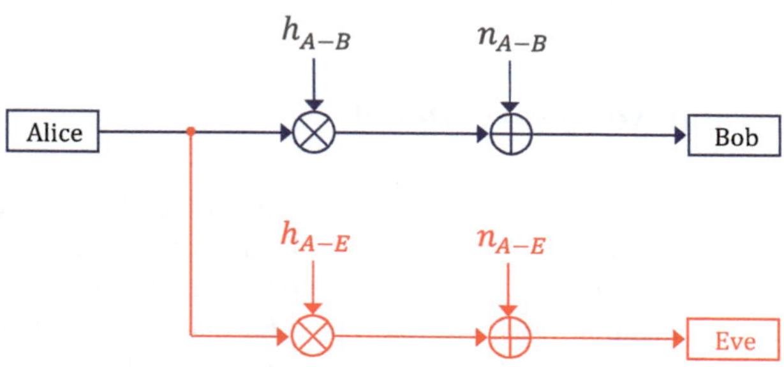

Fig. 6.2 The PLC wiretap channel model [178]

where P_T is the transmit power, $H_{A-X} = \|h_{A-X}\|^2$, and N_X is the relevant noise power.

The considered PLC channels are assumed to be Rayleigh fading channels [1, 7, 151]. Hence, the Probability Distribution Function (PDF) of H_{A-X} is expressed as follows:

$$f_{H_{A-X}}(x) = \frac{1}{\bar{H}_{A-X}(F_C, d_{A-X})} \exp\left(\frac{-x}{\bar{H}_{A-X}(F_C, d_{A-X})}\right), \tag{6.2}$$

where $\bar{H}_{A-X}(F_C, d_{A-X})$ is the mean of H_{A-X}, F_C is the used carrier frequency, and d_{A-X} is the power line length, which presents the transmission distance of the $A-X$ link.

Based on [7], the expression of $\bar{H}_{A-X}(F_C, d_{A-X})$ is given by

$$\bar{H}_{A-X}(F_C, d_{A-X}) = \exp\Big(-2\left[a_0 + a_1\, F_C^{a_2}\right] d_{A-X}\Big), \tag{6.3}$$

where a_0, a_1, and a_2 are the corresponding PLC environment's attenuation parameters.

Now, referring to the provided adversary system model, the following research concerns arise:

- In the presence of a malicious device, how secure is a PLC system at the physical layer level? In particular, how much information and data can be revealed to Eve?
- What effect does it have on the Secrecy Outage Probability (SOP) and the Average Secrecy Capacity (ASC)?

In the next sections, we look into the issues raised above by deriving the corresponding expressions of SOP and ASC, the two main metrics for PLS performance analysis that are often used in the literature. Furthermore, a new PLC performance analysis metric, named Secure and Successful Transmission Probability (SSTP), has been introduced and detailed, where we have derived its closed-form expression, according to the presented PLC system model and in two different scenarios with and without a friendly jamming technique.

6.2 Average Secrecy Capacity

The instantaneous secrecy capacity can be defined as the difference value between the instantaneous ergodic capacities of the main and the wiretap channels, i.e., $SC = EC_{A-B} - EC_{A-E}$. Therefore, the expression of the average secrecy capacity can be derived as follows:

$$ASC = \int_0^{\infty}\int_0^{\infty}\Big[\log_2\left(1+\gamma_{A-B}\right) - \log_2\left(1+\gamma_{A-E}\right)\Big] f_{H_{A-B}}(x) f_{H_{A-E}}(y) dx\, dy. \tag{6.4}$$

Using the expressions (6.1) and (6.2), the ASC expression can be rewritten as follows:

$$\begin{aligned} ASC = {} & \frac{1}{\bar{H}_{A-B}(F_C, d_{A-B})}\int_0^{\infty}\log_2\left(1+\frac{P_T\, x}{N_B}\right)\exp\left(\frac{-x}{\bar{H}_{A-B}(F_C, d_{A-B})}\right)dx \\ & - \frac{1}{\bar{H}_{A-E}(F_C, d_{A-E})}\int_0^{\infty}\log_2\left(1+\frac{P_T\, y}{N_E}\right)\exp\left(\frac{-y}{\bar{H}_{A-E}(F_C, d_{A-E})}\right)dy. \end{aligned} \tag{6.5}$$

Finally, after evaluating the integrations in (6.5), using [Eq. (4.337.2), [74]], the final ASC expression is given by

$$ASC = \frac{1}{\ln(2)}\left[\exp\left(\frac{-N_E}{P_T\ \bar{H}_{A-E}(F_C, d_{A-E})}\right)\mathrm{Ei}\left(\frac{-N_E}{P_T\ \bar{H}_{A-E}(F_C, d_{A-E})}\right)\right.$$
$$\left. - \exp\left(\frac{N_B}{P_T\ \bar{H}_{A-B}(F_C, d_{A-B})}\right)\mathrm{Ei}\left(\frac{-N_B}{P_T\ \bar{H}_{A-B}(F_C, d_{A-B})}\right)\right], \quad (6.6)$$

where Ei(.) is the exponential integral [74]

$$\mathrm{Ei}(-x) = -\int_{A-X}^{\infty} t^{-1} \exp(-t)\, dt. \quad (6.7)$$

To confirm and investigate the analytical results of the considered system model's ASC, we present the following numerical results and discussion.

Without losing generalizability, the used simulation parameters are selected to be $P_T = 10$ dBm, $N_B = N_E = -110$ dBm, $d_{A-B} = 20$ m, $d_{A-E} = [10 : 20]$ m, $SR_{Th} = 1$ b/s/Hz, $F_C \in \{10, 20, 50\}$ MHz, $a_0 = 0$, $a_1 = 7.8 \times 10^{-10}$, and $a_2 = 1$ (a_0, a_1, and a_2 are selected as per [130]).

Figure 6.3 presents the variations of ASC versus d_{A-E}, with different values of F_C. As shown in this figure, the ASC increases with the increased values of d_{A-E}, which is expected. This is due to the fact that an increase of the distance d_{A-E} results in a decrease of the received SNR at E and the corresponding ergodic capacity; hence, an increase of the ASC can be observed.

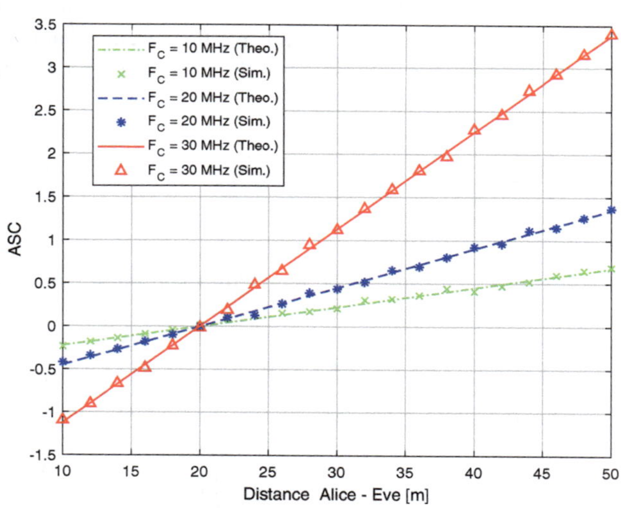

Fig. 6.3 The variations of ASC versus d_{A-E}, with different carrier frequencies [61]

Moreover, for $d_{A-E} < 20$ m (the distance of $A - B$ link), the ASC degrades significantly as the carrier frequency increases. This is explained by the fact that if $d_{A-E} < d_{A-B}$ and as shown in the channel attenuation formula in (6.3), increasing the carrier frequency has a negative influence on the received signal intensity at B when compared to that at E. In this situation, the channel attenuation has a greater influence on B than on E, which causes an impairment in the received SNR and the associated EC, resulting in a decrease of the ASC.

On the contrary, if E is far away from A, raising the carrier frequency has a significant negative impact on E, which in turn causes a drop in the relevant ergodic capacity. Hence, an increase in the ASC can be observed.

6.3 Secrecy Outage Probability

The SOP of a transmission link can be defined as the probability of the event when the corresponding secrecy capacity is less than a secrecy rate threshold SR_{Th} [176].

Consequently, the SOP can be expressed as follows:

$$SOP = Pr\left\{SC < SR_{\text{Th}}\right\}. \tag{6.8}$$

Furthermore, using the instantaneous secrecy capacity expression, (6.8) becomes

$$\begin{aligned} SOP &= Pr\left\{\left[\log_2(1+\gamma_{A-B}) - \log_2(1+\gamma_{A-E})\right] < SR_{\text{Th}}\right\} \\ &= Pr\left\{\left[\frac{1+\gamma_{A-B}}{1+\gamma_{A-E}}\right] < 2^{SR_{\text{Th}}}\right\} \\ &= Pr\left\{\gamma_{A-B} < \left(2^{SR_{\text{Th}}}\left[1+\gamma_{A-E}\right]-1\right)\right\}. \end{aligned} \tag{6.9}$$

By using the SNR expression in (6.1), the SOP expression can be rewritten as follows:

$$\begin{aligned} SOP &= Pr\left\{\frac{P_T\, H_{A-B}}{N_B} < \left(2^{SR_{\text{Th}}}\left[1+\frac{P_T\, H_{A-E}}{N_E}\right]-1\right)\right\} \\ &= Pr\left\{H_{A-B} < \left(2^{SR_{\text{Th}}}\left[\frac{N_B}{P_T}+\frac{N_B\, H_{A-E}}{N_E}\right]-\frac{N_B}{P_T}\right)\right\} \\ &= \int_0^{\infty}\int_0^{2^{SR_{\text{Th}}}\left[\frac{N_B}{P_T}+\frac{N_B\, z}{N_E}\right]-\frac{N_B}{P_T}} f_{H_{A-B}}(y)\, f_{H_{A-E}}(z)\, dy\, dz. \end{aligned} \tag{6.10}$$

By substituting the PDF expression (6.2) in (6.10) and after evaluating the first integration with respect to y, the SOP is expressed as

$$SOP = \int_0^\infty \left[1 - \exp\left(-\frac{2^{SR_{\text{Th}}}\left[\frac{N_B}{P_T} + \frac{N_B\, z}{N_E}\right] - \frac{N_B}{P_T}}{\bar{H}_{A-B}(F_C, d_{A-B})} \right) \right] f_{H_{A-E}}(z)\, dz$$

$$= 1 - \exp\left(-\frac{N_B[2^{SR_{\text{Th}}} - 1]}{P_T\, \bar{H}_{A-B}(F_C, d_{A-B})} \right)$$

$$\times \int_0^\infty \exp\left(-\frac{2^{SR_{\text{Th}}} N_B\, z}{N_E \bar{H}_{A-B}(F_C, d_{A-B})} \right) f_{H_{A-E}}(z)\ dz. \tag{6.11}$$

Now, by using the PDF expression of H_{A-E}, the SOP expression can be represented as follows:

$$SOP = 1 - \exp\left(-\frac{N_B[2^{SR_{\text{Th}}} - 1]}{P_T\, \bar{H}_{A-B}(F_C, d_{A-B})} \right)$$

$$\times \int_0^\infty \exp\left(-\frac{2^{SR_{\text{Th}}} N_B\, z}{N_E \bar{H}_{A-B}(F_C, d_{A-B})} \right) f_{H_{A-E}}(z) dz$$

$$= 1 - \exp\left(-\frac{N_B[2^{SR_{\text{Th}}} - 1]}{P_T\, \bar{H}_{A-B}(F_C, d_{A-B})} \right) \frac{1}{\bar{H}_{A-E}(F_C, d_{A-E})}$$

$$\times \int_0^\infty \exp\left(-z\left[\frac{2^{SR_{\text{Th}}} N_B}{N_E \bar{H}_{A-B}(F_C, d_{A-B})} + \frac{1}{\bar{H}_{A-E}(F_C, d_{A-E})} \right] \right) dz. \tag{6.12}$$

After evaluating the integration, followed by some simplifications, the final SOP expression is given by

$$SOP = 1 - \frac{N_E \bar{H}_{A-B}(F_C, d_{A-B})}{2^{SR_{\text{Th}}} N_B \bar{H}_{A-E}(F_C, d_{A-E}) + N_E \bar{H}_{A-B}(F_C, d_{A-B})}$$

$$\times \exp\left(-\frac{N_B[2^{SR_{\text{Th}}} - 1]}{P_T\, \bar{H}_{A-B}(F_C, d_{A-B})} \right). \tag{6.13}$$

The derived SOP expression in (6.13) has been used to investigate and analyze the transmission security performance of the PLC adversary system model under

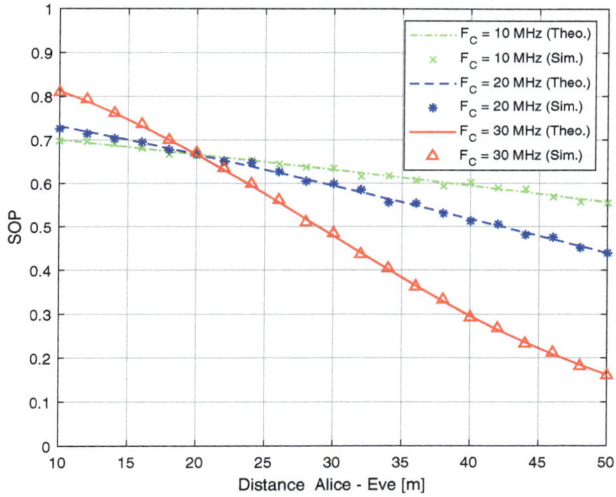

Fig. 6.4 The variations of SOP versus d_{A-E}, with different carrier frequencies [61]

consideration. Accordingly, Fig. 6.4 presents the corresponding SOP variations versus different values of d_{A-E}. In these simulation results, the same parameter values for ASC have been used.

As presented in this figure, the SOP variation is inversely proportional to the transmission distance of the $A - E$ link, which is expected. Indeed, increasing the d_{A-E} values decreases the SNR at Eve, which results in an increase of the secrecy capacity, and hence, a decrease of the relevant SOP can be observed. In addition, as previously explained for the ASC simulation results, decreasing the carrier frequency is recommended to decrease the SOP when Eve is near to $Alice$. And, in the contrary scenario, when Eve is far away from $Alice$, increasing F_C has a significant impact on Eve's SNR, which results in a clear enhancement of the system's performance scored by the SOP metric.

6.4 Secure and Successful Transmission Probability

The SSTP metric has been introduced to evaluate both the security and reliability performances of a given transmission link. Accordingly, it can be defined as the probability of the event when the corresponding secrecy capacity is larger than a given secrecy rate threshold (SR_{Th}), and the SNR value is greater than a given SNR threshold (γ_{Th}). Consequently, the SSTP can be expressed as follows:

$$SSTP = \Pr\Big\{\big(SC \geq SR_{\text{Th}}\big) \cap \big(\gamma_{A-B} \geq \gamma_{\text{Th}}\big)\Big\}. \tag{6.14}$$

The derivations of the SSTP expressions, with and without a friendly jamming scenario, are detailed below.

6.4.1 Derivation of the SSTP Expression Without a Jamming Technique

In this scenario, we consider the presented adversary system model without the employment of a jamming technique.

By using the secrecy capacity definition, the SSTP expression can be derived as follows:

$$\begin{aligned} SSTP &= \Pr\Big\{\Big(\big[\log_2(1+\gamma_{A-B}) - \log_2(1+\gamma_{A-E})\big] \geq SR_{\text{Th}}\Big) \cap \big(\gamma_{A-B} \geq \gamma_{\text{Th}}\big)\Big\} \\ &= \Pr\left\{\left(\left[\frac{1+\gamma_{A-B}}{1+\gamma_{A-E}}\right] \geq 2^{SR_{\text{Th}}}\right) \cap \left(\gamma_{A-B} \geq \gamma_{\text{Th}}\right)\right\} \\ &= \Pr\left\{\left(\gamma_{A-B} \geq \left(2^{SR_{\text{Th}}}\left[1+\gamma_{A-E}\right] - 1\right)\right) \cap \left(\gamma_{A-B} \geq \gamma_{\text{Th}}\right)\right\}. \end{aligned} \tag{6.15}$$

Furthermore, it can be rewritten as follows:

$$\begin{aligned} SSTP &= \Pr\left\{\gamma_{A-B} \geq \max\left(\Gamma_{\text{Th}}, \gamma_{\text{Th}}\right)\right\} \\ &= \begin{cases} \Pr\left\{\gamma_{A-B} \geq \Gamma_{\text{Th}}\right\}, & \text{if } (\Gamma_{\text{Th}} \geq \gamma_{\text{Th}}) \\ \Pr\left\{\gamma_{A-B} \geq \gamma_{\text{Th}}\right\}, & \text{if } (\Gamma_{\text{Th}} < \gamma_{\text{Th}}) \end{cases} \end{aligned} \tag{6.16}$$

where $\Gamma_{\text{Th}} = \left(2^{SR_{\text{Th}}}\left[1+\gamma_{A-E}\right] - 1\right)$.

Accordingly, the $SSTP$ expression is derived as follows:

$$SSTP = \int_{\gamma_{th}}^{\infty} f_{\Gamma_{th}}(y) \int_{y}^{\infty} f_{\gamma_{A-B}}(x)\, dx\, dy + \int_{0}^{\gamma_{th}} f_{\Gamma_{th}}(y) \int_{\gamma_{th}}^{\infty} f_{\gamma_{A-B}}(x)\, dx\, dy, \tag{6.17}$$

where $f_{\gamma_{A-B}}$ and $f_{\Gamma_{th}}$ are the PDF expressions of γ_{A-B} and Γ_{th}, respectively.

In this chapter, we assume that the main and the wiretap channels are Rayleigh distributed, where the PDF expressions are given by [174]

$$f_{\gamma_{A-B}}(x) = \frac{1}{\bar{\gamma}_{A-B}} \exp\left(\frac{-x}{\bar{\gamma}_{A-B}}\right), \tag{6.18}$$

and

$$f_{\Gamma_{th}}(y) = \frac{1}{\bar{\Gamma}_{th}} \exp\left(\frac{-y}{\bar{\Gamma}_{th}}\right), \tag{6.19}$$

respectively, where $\bar{\gamma}_{A-B}$ is the mean of γ_{A-X}, which is expressed as

$$\bar{\gamma}_{A-B} = \frac{P_{\mathrm{T}}\,\bar{H}_{A-B}}{N_B}, \tag{6.20}$$

and $\bar{\Gamma}_{th}$ is the mean of Γ_{th}, which is expressed as

$$\bar{\Gamma}_{th} = 2^{SR_{\mathrm{Th}}}\left[1 + \frac{P_{\mathrm{T}}\,\bar{H}_{A-E}}{N_E}\right] - 1. \tag{6.21}$$

After evaluating the first two integrations, using the corresponding PDF expressions of γ_{A-E} and Γ_{th}, (6.17) yields to

$$\begin{aligned} SSTP = {} & \frac{\bar{\gamma}_{A-B}}{\bar{\gamma}_{A-B} + \bar{\Gamma}_{th}} \exp\left(- \gamma_{th}\left[\frac{1}{\bar{\Gamma}_{th}} + \frac{1}{\bar{\gamma}_{A-B}}\right]\right) \\ & + \int_0^{\gamma_{th}} f_{\Gamma_{th}}(y) \int_{\gamma_{th}}^{\infty} f_{\gamma_{A-B}}(x)\, dx\, dy. \end{aligned} \tag{6.22}$$

By evaluating the integrations in (6.22), with respect to x and y, followed by some simplifications, the final SSTP expression is given by

$$\begin{aligned} SSTP = {} & \frac{\bar{\gamma}_{A-B}}{\bar{\gamma}_{A-B} + \bar{\Gamma}_{th}} \exp\left(- \gamma_{th}\left[\frac{1}{\bar{\Gamma}_{th}} + \frac{1}{\bar{\gamma}_{A-B}}\right]\right) \\ & + \left[1 - \exp\left(\frac{-\gamma_{th}}{\bar{\Gamma}_{th}}\right)\right] \exp\left(\frac{-\gamma_{th}}{\bar{\gamma}_{A-B}}\right). \end{aligned} \tag{6.23}$$

6.4.2 Derivation of the SSTP Expression with a Jamming Technique

In this subsection, we detail the derivation of the SSTP expression for the considered adversary system model, where a friendly jamming technique is deployed. In this scenario, the jamming technique consists of injecting artificial noise to avoid data leakage to malicious PLC devices when *Alice* communicates with *Bob*. In particular, *Alice* should generate and inject an artificial noise signal in parallel with the useful data transmission. This generated noise signal should be shared with *Bob* to filter the received signal and detect the useful data only, which is not the case on the eavesdropper's side.

Accordingly, using (6.14)–(6.16), the corresponding $SSTP$ expression is derived as follows:

$$SSTP^{\mathrm{FJ}} = \Pr\left\{\gamma_{A-B} \geq \max\left(\Gamma^{\mathrm{eq}}_{\mathrm{Th}}, \gamma_{\mathrm{Th}}\right)\right\}$$
$$= \begin{cases} \Pr\left\{\gamma_{A-B} \geq \Gamma^{\mathrm{eq}}_{\mathrm{Th}}\right\}, & \text{if } (\Gamma^{\mathrm{eq}}_{\mathrm{Th}} \geq \gamma_{\mathrm{Th}}), \\ \Pr\left\{\gamma_{A-B} \geq \gamma_{\mathrm{Th}}\right\}, & \text{if } (\Gamma^{\mathrm{eq}}_{\mathrm{Th}} < \gamma_{\mathrm{Th}}), \end{cases} \tag{6.24}$$

where $\Gamma^{\mathrm{eq}}_{\mathrm{Th}} = 2^{SR_{\mathrm{Th}}}\left[1 + \gamma^{\mathrm{eq}}_{A-E}\right] - 1$.

Here, $\gamma^{\mathrm{eq}}_{A-E}$ presents the equivalent instantaneous SNR of the wiretap channel, which is expressed as follows:

$$\gamma^{\mathrm{eq}}_{A-E} = \frac{P_{\mathrm{T}}\, H_{A-E}}{P_{\mathrm{FJ}}\, H_{A-E} + N_E}, \tag{6.25}$$

where P_{FJ} denotes the artificial noise signal power.

Consequently, the $SSTP^{\mathrm{FJ}}$ can be rewritten as follows:

$$SSTP^{\mathrm{FJ}} = \begin{cases} \Pr\left\{\gamma_{A-B} \geq \Gamma^{\mathrm{eq}}_{\mathrm{Th}}\right\}, & \text{if } (H_{A-E} \geq \beta), \\ \Pr\left\{\gamma_{A-B} \geq \gamma_{\mathrm{Th}}\right\}, & \text{if } (H_{A-E} < \beta), \end{cases} \tag{6.26}$$

where

$$\beta = \frac{N_E\left(1 + \gamma_{\mathrm{Th}} - 2^{SR_{\mathrm{Th}}}\right)}{P_{\mathrm{T}}\, 2^{SR_{\mathrm{Th}}} - P_{\mathrm{FJ}}\left(1 + \gamma_{\mathrm{Th}} - 2^{SR_{\mathrm{Th}}}\right)}. \tag{6.27}$$

By using the PDF expressions of γ_{A-B} and H_{A-E}, (6.26) is derived as follows:

$$SSTP^{\mathrm{FJ}} = \int_{\beta}^{\infty} f_{H_{A-E}}(y) \int_{\Gamma^{\mathrm{eq}}_{\mathrm{Th}}}^{\infty} f_{\gamma_{A-B}}(x)\, dx dy + \int_{0}^{\beta} f_{H_{A-E}}(y) \int_{\gamma_{\mathrm{Th}}}^{\infty} f_{\gamma_{A-B}}(x)\, dx dy.$$
$$= \int_{\beta}^{\infty} \frac{1}{\bar{H}_{A-E}} \exp\left(\frac{-y}{\bar{H}_{A-E}}\right) \exp\left(\frac{-\Gamma^{\mathrm{eq}}_{\mathrm{Th}}}{\bar{\gamma}_{A-B}}\right) dy$$
$$+ \left[1 - \exp\left(\frac{-\beta}{\bar{H}_{A-E}}\right)\right] \exp\left(\frac{-\gamma_{\mathrm{Th}}}{\bar{\gamma}_{A-B}}\right). \tag{6.28}$$

Table 6.1 The simulation parameters [178]

Parameter	Value
Number of iterations	10^4
P_{T} [dBm]	1
N_B [dBm]	−30
d_{A-B} [m]	50
d_{A-E} [m]	[10 : 100]
SR_{Th} [b/s/Hz]	1
γ_{Th}	2
F_C [MHz]	10, 20, and 50
a_0	0
a_1	7.8 10^{-10}
a_2	1

Finally, using the change in variables, $x = \frac{y-\beta}{\bar{H}_{A-E}}$, and the Laguerre theorem, the final $SSTP^{\mathrm{FJ}}$ expression is given by

$$SSTP^{\mathrm{FJ}} = \sum_{n=1}^{N}\left\{w_n \exp\left(-\frac{1}{\bar{\gamma}_{A-B}}\left[2^{SR_{\mathrm{Th}}}\left(1+\frac{P_{\mathrm{T}}(x_n\bar{H}_{A-E}+\beta)}{P_{\mathrm{FJ}}(x_n\bar{H}_{A-E}+\beta)+N_E}\right)-1\right]\right)\right\}$$
$$\times \exp\left(\frac{-\beta}{\bar{H}_{A-E}}\right)+\left[1-\exp\left(\frac{-\beta}{\bar{H}_{A-E}}\right)\right]\exp\left(\frac{-\gamma_{\mathrm{Th}}}{\bar{\gamma}_{A-B}}\right). \tag{6.29}$$

where x_n, w_n, and N are the roots, weights, and the order of the Laguerre polynomial, respectively [175].

The remainder of this chapter presents and investigates the considered system performances in terms of security and link reliability. To this end, we have conducted Monte Carlo simulations of the friendly jamming technique within the presented PLC adversary system model.

Without loss of generality, the used simulation parameters' values are presented in Table 6.1.

Figure 6.5 presents the variations of $SSTP$ versus d_{A-E}, without the use of a friendly jamming technique and with different carrier frequencies. The results show that the SSTP increases with the increased distance d_{A-E}, which is logical. Indeed, increasing this transmission distance decreases Eve's SNR, which results in a decrease of the relevant ergodic capacity, and hence, an increase of the SSTP can be observed. Additionally, it is shown that for $d_{A-E} < 50$ m (the $A-B$ distance), the increase of the used carrier frequency significantly decreases the SSTP. This is because, when $d_{A-E} < d_{A-B}$, an increase of F_C has a negative impact on the received signal strength at B when compared to that at E, i.e., the channel attenuation effect impacts more B, which results in a decrease of the corresponding $SSTP$. On the contrary, if Eve is far away from $Alice$, the increase in carrier frequency

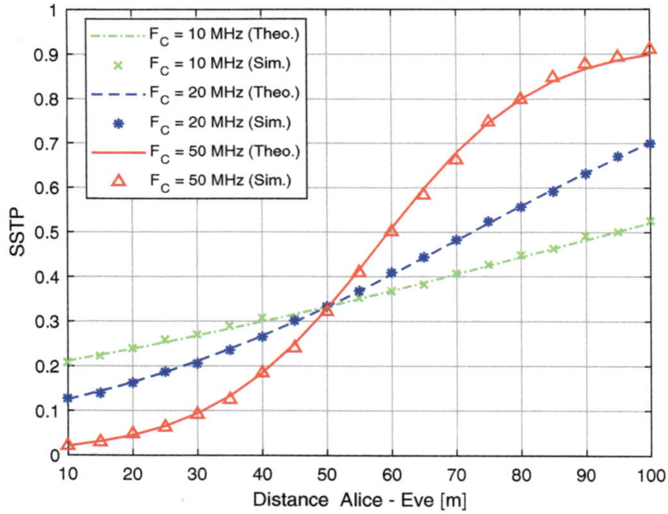

Fig. 6.5 The variations of $SSTP$ versus d_{A-E}, with different carrier frequencies and without a friendly jamming technique [178]

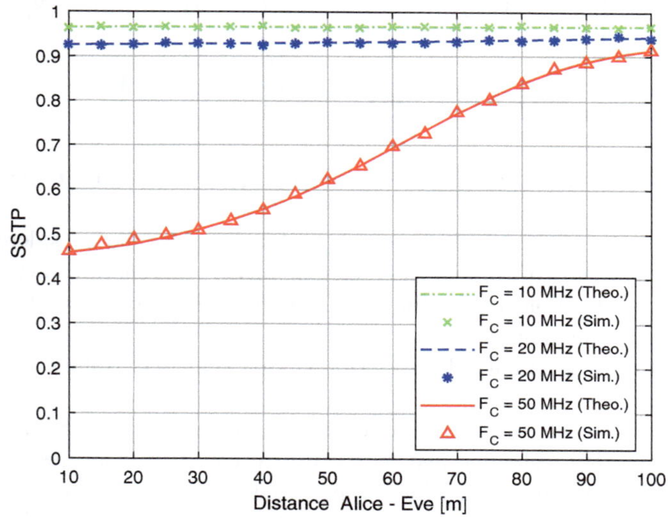

Fig. 6.6 The variations of $SSTP$ versus d_{A-E}, with a friendly jamming technique, $P_{FJ} = 0.1P_T$, and different carrier frequencies [178]

significantly decreases Eve's ergodic capacity, and hence, an enhancement of the SSTP can be observed.

To investigate the impact of the friendly jamming technique on the considered adversary system model, we present in Fig. 6.6 the SSTP variations versus d_{A-E},

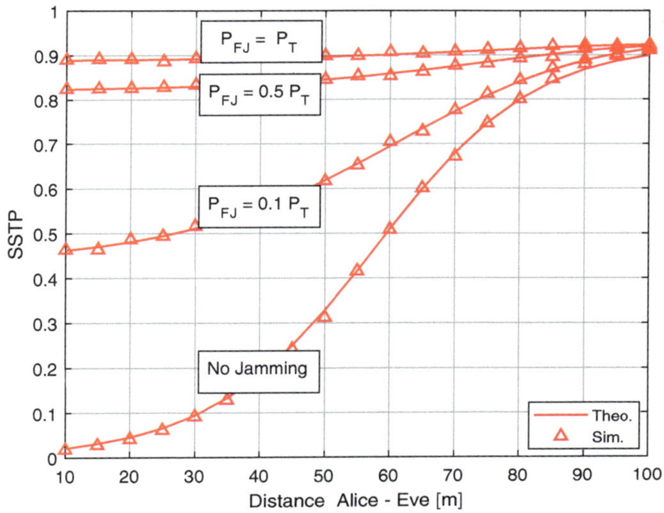

Fig. 6.7 The variations of $SSTP$ versus d_{A-E}, with different friendly jamming transmit power [178]

with a jamming signal power $P_{FJ} = 0.1P_T$, and different carrier frequencies. As shown in this figure, for the different used carrier frequencies, a clear enhancement can be observed, which is expected. This is due to the fact that the main idea of the friendly jamming technique consists of increasing the received noise at Eve, which significantly decreases the corresponding SNR, and hence, the SSTP increases.

Figure 6.7 presents the impact of the jamming signal power level on the SSTP variations. In particular, for a short distance d_{A-E}, increasing the jamming signal power increases the relevant SSTP, which is not the case for long transmission distances. This can be explained by the fact that the additive noise has more impact on Eve's SNR than that of the channel attenuation when Eve is near to $Alice$. However, if Eve is far away from $Alice$, the strength of the received signal at Eve and the relevant SNR will significantly be decreased due to the channel attenuation, which will increase the associated SSTP. As a result, increasing the jamming signal power has a minor effect on the corresponding SSTP. In this case, where Eve is far away from the legitimate transmitter $Alice$, friendly jamming can be considered a useless precaution technique.

Chapter 7
Confidentiality Technique for PLC Networks

Abstract This chapter presents and details a confidentiality technique for Power Line Communication (PLC) networks. It consists of a new multilevel quantization algorithm suitable for key generation, which facilitates the application of information reconciliation strategies to mitigate the impact of channel estimation errors. The method addresses the general lack of channel symmetry in the power line by utilizing the path delays of the Channel Impulse Response (CIR) to derive a symmetric virtual CIR that can lead to a decrease in the Bit Mismatch Rate (BMR) by at least three orders of magnitude.

7.1 PLC Channel Modeling

Before introducing the novel Physical Layer Security (PLS) key generation method, we examine the relationship between the Power Line Communication (PLC) topology and the relevant channel characteristics to demonstrate the PLC channel symmetry dependence on path delays. To this end, we start by describing the reciprocity characteristics of PLC channels, which present the fundamental concept of PLS key generation. Then, we focus on the deterministic and the empirical modeling methods that have been considered as the main PLC channel modeling approaches in the literature [43, 130, 149, 262].

7.1.1 PLC Channel Reciprocity and PLS Key Generation

The PLC and the wireless communication channels differ considerably. In fact, the variations of the wireless Channel State Information (CSI) at the transmitters and their receivers are highly correlated [239, 258], which is not the case for PLC, due to the impedance imbalance [130, 186]. In particular, the wireless devices are set to operate at 50 ohms by default. However, this is not the case for PLC, where the traditional PLC transmitters and receivers are configured to enhance the

J. Hernandez Fernandez et al., *Physical Layer Security in Power Line Communications*, Advances in Information Security 108,
https://doi.org/10.1007/978-3-031-57349-1_7

transferred voltage, using different impedances, which results in a lack of reciprocity [186]. Nevertheless, the CIRs at a given PLC transmitter and its receiver are highly correlated in terms of the corresponding path delay [186, 258].

7.1.2 Deterministic Modeling Approach

This method is founded on the Transmission Line (TL) theory, which calls for acquiring the two-port network-based ABCD matrix that defines the correlation between input and output voltages and currents [43, 149]. This connection can be presented as follows:

$$H = \begin{bmatrix} V_1 \\ I_1 \end{bmatrix} \begin{bmatrix} A & B \\ C & D \end{bmatrix} \begin{bmatrix} V_2 \\ I_2 \end{bmatrix}, \tag{7.1}$$

where $V_1(I_1)$ is the input voltage (current) and $V_2(I_2)$ is the output voltage (current), $A = D = \cosh(\gamma l)$, $B = Z_c \sinh(\gamma l)$, $C = \sinh(\gamma l)/Z_c$, where l is the corresponding conductor's length, Z_c is the characteristic impedance, and γ is the signal propagation parameter, where its expression is given by Chu et al. [43], Masood et al. [149]

$$\gamma = \sqrt{(R + j\omega L)(G + j\omega C)}, \tag{7.2}$$

here, R, L, C and G are the Power Line (PL) parameters, and ω is the angular frequency. Furthermore, the TL theory-based transfer function can be expressed as follows:

$$H = 20 \log_{10} \left(\frac{Z_L}{A\, Z_L + B + C\, Z_L\, Z_S + D\, Z_S} \right), \tag{7.3}$$

where Z_S and Z_L represent the source and load impedances.

It is worth mentioning that this deterministic approach has a high level of complexity and restricted viability since it requires in-depth knowledge of the installed transmission lines and the relevant power network parameters.

7.1.3 Empirical Modeling Approach

This approach is a time domain-based modeling technique proposed for the first time by Zimmerman in [262]. It consists of a statistical model that is based on the power line channel's multipath properties. In accordance with this model, the PLC CIR is expressed as follows [130, 262]:

$$h(t) = \sum_{i=0}^{N-1} k_i \, \delta(t - \tau_i), \tag{7.4}$$

where N is the number of the significant (nonnegligible) paths, and k_i and τ_i are the attenuation coefficient and the delay of the ith path, respectively. An N-tap Finite Impulse Response (FIR) filter can be used to implement this channel model, where the delay τ_0 presents the first (direct) path and k_0 is the relevant attenuation. All the following tap indexes i, $i \in \{1, .., N\}$, present the echoes of the first one.

In view of this, we detail in the following part the corresponding Channel Frequency Response (CFR) expression, where we confirm the independence of the PLC channel multipath delays, τ_i, from the PL characteristics while depending only on the PL network topology of the communicated nodes.

By definition, evaluating the Fourier transform of the CIR expression in (7.4) yields to the following CFR expression:

$$H(f) = \sum_{i=1}^{N} k_i \, \exp(-j2\pi f \tau_i), \tag{7.5}$$

where f presents the carrier frequency. In practical settings, the frequency and the used cable length have a direct impact on the values of the attenuation coefficient k_i, which can be expressed as follows [130, 262]:

$$k_i = g_i \, \exp(-\alpha(f) \, l_i). \tag{7.6}$$

This expression is the analysis result of different power line channel measurements, where $\alpha(f)$ presents the frequency-dependent attenuation coefficient, l_i denotes the considered cable length (length of path index i), and g_i is the relevant weight factor that presents the PLC signal propagation and reflection effects.

Consequently, using the expressions (7.5) and (7.6), the corresponding CFR expression can be written as follows:

$$H(f) = \sum_{i=0}^{N-1} g_i \, \exp(-\alpha(f) \, l_i) \, \exp\left(-j2\pi f \tau_i\right). \tag{7.7}$$

The attenuation factor $\alpha(f)$ in this equation was evaluated and analyzed for different types of power lines, which yields to presenting it as follows [262]:

$$\alpha(f) = \frac{R^{'}}{2Z_c} = a_0 + a_1 f^{a_2}, \tag{7.8}$$

where $R^{'}$ presents the resistance per length unit, and a_0, a_1, and a_2 are power line propagation parameters that depend on the characteristic impedance Z_c and the power line characteristics [130, 262]. In view of this, the empirical modeling-based expression of the PLC CFR is given by

$$H(f) = \sum_{i=0}^{N-1} g_i \exp\left(- \left[a_0 + a_1 f^{a_2} + j \frac{2\pi f}{v_p} \right] l_i \right), \tag{7.9}$$

where $v_p = \frac{l_i}{\tau_i}$, which presents the PLC signal propagation speed.

From the presented CFR expression in (7.7), it is worth noting that the path gain k_i depends on the power line impedance and characteristics, while the path delays are related to the power line topology only.

In the literature, many works have evaluated and verified the presented empirical modeling approach that has shown good agreement with relevant PLC channel measurements on a statistical basis [130].

7.2 PLS Key Generation Scheme

In this section, we present the proposed PLS key generation scheme, where we detail the corresponding steps from the channel probing to the reconciliation process.

7.2.1 Channel Probing

The channel probing process involves the estimation of the considered CIRs from the received Orthogonal Frequency Division Multiplexing (OFDM) signals in the frequency domain. This estimation consists of two main processes. First, using predefined signal pilots and a given interpolation method, the CFR can be estimated in the frequency domain. Then, using the Inverse Discrete Fourier Transform (IDFT), the CFR is converted into the CIR.

7.2.2 Quantization

As detailed in previous sections, PLC channels are asymmetric except for channel path delays, which are reciprocal. As a consequence, symmetric keys cannot be generated from the CIR using a direct multilevel quantization since the peaks will differ in amplitude. As a result, the multilevel quantization techniques and their advantages have not been utilized.

In this work, we have addressed the cited constraint by transforming the asymmetric CIR to a symmetric virtual CIR. Following that, we use our proposed Algorithm 1 to conduct a multilevel quantization. The output of this algorithm is a

Algorithm 1 Multilevel quantization [87]

```
Inputs: Key Length: k_Len.
        CIR Estimates: CIRe.
        One level Quantization Threshold: L1.
        Multilevel Quantization Thresholds:
        L- = -1 and L+ = +1.
        Multilevel Quantization Step: α.
        m = 1.
Step 1: CIRe Magnitude One level Quantization:
for n from 1 to k_Len do
    if (|CIRe(n)| ≤ L1) then
        k0(n) = 1
    else
        k0(n) = 0
    end if
end for
Step 2: Virtual CIR (VCIR) Generation:
Part1 = k0(1 : k_Len/2)
Part2 = k0(k_Len/2 + 1 : k_Len)
Generating the VCIR, according to Table I.
Step 3: VCIR Multilevel Quantization
while (m < k_Len/2) and (L+ ≠ L-) do
    for n from 1 to k_Len/2 do
        if (VCIR(n) ≥ L+) then
            key(m) = 1
            Pos(m) = n
            m = m + 1
        end if
        if (VCIR(n) ≤ L-) then
            key(m) = 0
            Pos(m) = n
            m = m + 1
        end if
    end for
    L+ = (L+) - α
    L- = (L-) + α
end while
```

unique key that is generated by the two communicating devices. More specifically, this key generation method requires the following five parameters as inputs:

1. The estimated CIR: $CIRe$
2. The key length: k_{Len}
3. One-level quantization threshold $L1$, which is a parameter related to the average noise level—to prevent detecting false CIR peaks
4. The multilevel quantization thresholds, $L-$ and $L+$, initialized to $+1$ and -1, respectively
5. The multilevel quantization step: α

Table 7.1 The VCIR generation table for $n \in \{1, 2, \ldots, k_{Len}/2\}$ [87]

$Part1(n)$	$Part2(n)$	$VCIR(n)$
1	1	1
1	0	−1
0	1	-0.5 if $VCIR(n-1) \leq 0$
		$+0.5$ if $VCIR(n-1) > 0$
0	0	0

These inputs should be used by the proposed multilevel quantization Algorithm 1 to perform the following three main steps:

- **Step 1** consists of conducting a single-level quantization of the estimated CIR to generate an initial key $k0$ of length k_{Len}.
- **Step 2** splits the generated vector $k0$ into $Part1$ and $Part2$. These two parts are then converted to the relevant VCIR, using Table 7.1.
- Finally, in **Step 3**, the VCIR quantization should be conducted to generate the corresponding key and Pos vectors at each communicated node. In particular, as shown in Algorithm 1, the VCIR should be scanned to detect the samples that cross or equal the first quantization level $L-$ and $L+$. Then, this quantization process should be repeated, with an adjustment in the upper and lower levels, until the generated key length equals $k_{Len}/2$ or $L- = L+$. During this multilevel quantization, the position of each detected peak should be added to a position vector, noted by Pos, while its level should be converted to one or zero to generate the final key.

To further clarify the description of the proposed multilevel quantization algorithm, we illustrate in Fig. 7.1, a walk-through example for the abovementioned three steps. Without loss of generality and to simplify this explanation, a short key of 32 bits and four quantization levels are considered in this example.

7.2.3 Reconciliation

The reconciliation consists of a signaling procedure between the communicated nodes to make sure that the generated keys are the same. For instance, let us consider two nodes, A and B. By using Algorithm 1, the keys and the corresponding position vectors for both links, $(A \rightarrow B)$ and $(B \rightarrow A)$, can be generated. Then, A sends its position vector Pos_{B-A} to B. After that, B should compare Pos_{B-A} with its position vector Pos_{A-B} to identify the agreement bits and generate an updated position vector that will be sent back to A. Subsequently, both nodes A and B should locally update their keys key_{A-B} and key_{B-A} based on the updated position vector.

To further explain the reconciliation procedure, we consider a simulated PLC scenario with three possible cases that could occur during this procedure.

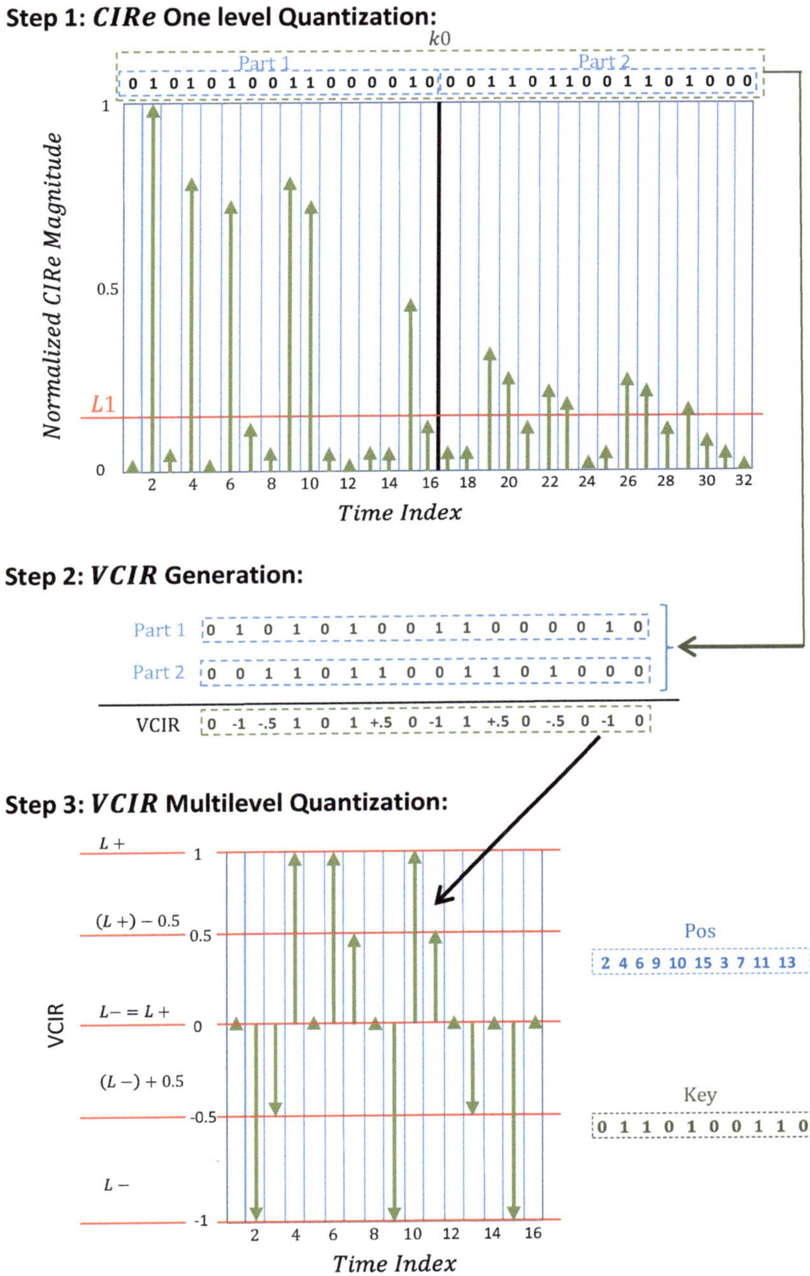

Fig. 7.1 Output examples of Algorithm 1 steps, with $k_{Len} = 32$, $L1 = 0.2$, $\alpha = 0.5$ [87]

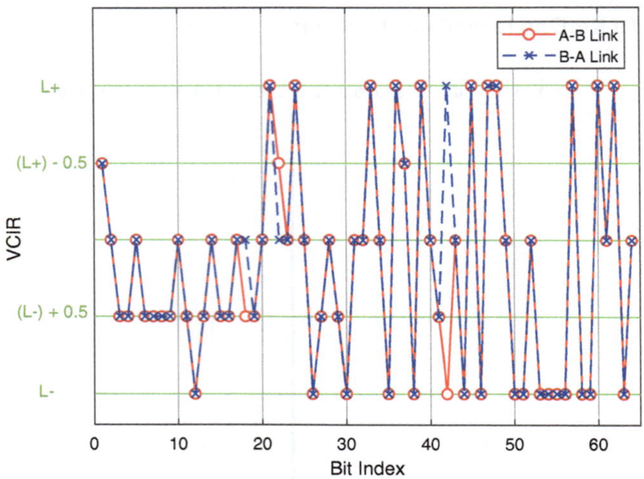

Fig. 7.2 Virtual channel impulse response variations [87]

To this end, we have developed a PLC simulator that is able to generate the CIRs and their relevant VCIRs for different PLC topologies and communication parameters. Further details about the developed simulator are presented in the next subsection, along with numerical results and discussion.

Accordingly, we consider the VCIRs presented in Fig. 7.2 that contains the following initial keys and position vectors for the links $(A \rightarrow B)$ and $(B \rightarrow A)$:

$$Key_{A-B} = [01100101010010110000001001101000000000000010010]$$

$$\begin{aligned} Pos_{A-B} = [&12\ 21\ 24\ 26\ 30\ 33\ 35\ 36\ 38\ 39\ 42\ 44\ 45\ 46\ 47\\ &48\ 50\ 51\ 53\ 54\ 55\ 56\ 57\ 58\ 59\ 60\ 62\ 63\ 1\ \ 3\\ &4\ \ 6\ 7\ 8\ 9\ 11\ 13\ 15\ 16\ 18\ 19\ 22\ 27\ 29\ 37\ 41] \end{aligned}$$

$$Key_{B-A} = [011001010110101100000010011010000000000000010]$$

$$\begin{aligned} Pos_{B-A} = [&12\ 21\ 24\ 26\ 30\ 33\ 35\ 36\ 38\ 39\ 42\ 44\ 45\ 46\ 47\\ &48\ 50\ 51\ 53\ 54\ 55\ 56\ 57\ 58\ 59\ 60\ 62\ 63\ 1\ \ 3\\ &4\ 6\ 7\ 8\ 9\ 11\ 13\ 15\ 16\ 19\ 27\ 29\ 37\ 41] \end{aligned}$$

As shown in Pos_{B-A} vector, peaks 18 and 22 (in red color) are not present on both sides. These peaks should be ignored when generating the final keys at A and B, which yields to

$$Key_{A-B} = [01100101010010110000001001101000000000000000010]$$

$$Key_{B-A} = [01100101011010110000001001101000000000000000010]$$

Despite being effective at removing peaks that are mismatched in terms of position, this signaling procedure is unable to prevent peaks that are at the same position but have different levels, e.g., the detected peak at position 42 (in blue color), which results in a bit mismatch between the keys Key_{A-B} and Key_{B-A}. In this case, some useful techniques can be deployed to guarantee the use of the same final key. One of these techniques has been presented in [19], where A can generate a random number R, encrypt it with its new key, and send the encrypted message to B. After that, using its key, B should decode the received message to get a number $\hat{R}$. Then, encrypt the number $(\hat{R} + 1)$ and send it to A. Finally, if A is able to encrypt the received message and get the number $(R + 1)$, then a key agreement acknowledgment will be sent to B. Otherwise, the key is discarded, and the key generation process should be restarted.

7.3 Results and Discussion

In this section, we analyze and validate the performances of the proposed PLS key generation scheme in different simulated PLC scenarios. First, the conducted simulation setup is described; then the simulation results are presented and discussed.

7.3.1 *Simulation Setup*

In order to assess the effectiveness and the performance of the proposed scheme, we have developed a generic PLC simulator that involves the following main functions:

1. **Topology Simulation**
   The development of a generic statistical model that would represent all topology configurations is a challenging task out of this chapter's scope. For this reason, we focus on three main topologies that present the most common medium and low-voltage scenarios: bus, tree, and star topologies. The star topology has been considered to analyze direct connections without discontinuities. The bus and tree topologies present the typical distribution networks, where end nodes are connected to the main power line via secondary lines.

   To generate the corresponding path delays and gains for each topology, the PLC time domain model introduced in [130] was used within the developed PLC simulator. This model describes the PLC signal multipath propagation phenomenon, when the signal can be reflected and refracted at any discontinuity, terminal node, or branch, resulting in a number of arriving paths between two given endpoints for each topology.

To further explain the implementation of the multipath propagation model within the developed simulator, we consider an example of a simple T scenario, with end nodes A, B, D, and C as the middle node connecting all the nodes. Let L_{XY} denote the generic distance between two given nodes X and Y, α the propagation attenuation coefficient per PL length unit, ρ_X the reflection attenuation coefficient at node X, and δ_X the discontinuity attenuation coefficient at node X.

Taking into consideration the abovementioned factors, the expression of the first path's length is given by

$$l_1 = L_{AC} + L_{CB} = L_{AB}, \tag{7.10}$$

and the relevant path's gain can be written as follows:

$$g_1 = (1 - \alpha L_{AC})\,(1 - \alpha L_{CB})\,(1 - \delta_C). \tag{7.11}$$

For the second path ($A \rightarrow C \rightarrow D \rightarrow C \rightarrow B$), the corresponding path's length and gain are given by

$$l_2 = L_{AC} + 2L_{CD} + L_{CB}, \tag{7.12}$$

and

$$g_2 = (1 - \alpha L_{AC})(1 - \alpha L_{CD})^2(1 - \alpha L_{CB})(1 - \rho_D)(1 - \delta_C)^2, \tag{7.13}$$

respectively. Following the same mechanism, during a predefined channel probing period, a set of arriving paths are evaluated by computing their path delays and gains. The corresponding CIR is then generated. After that, the CIR should be converted to the frequency domain to get the relevant CFR that presents the ideal CSI. This CFR will be used to generate the received signal. Sequentially, the least-squares estimation method is used in this simulator to estimate the CFR, where the channel estimation error is inversely proportional to the considered link's instantaneous complex noise value. Finally, the IDFT is used to generate the estimated CIR.

It is important to note that the multipath detection accuracy depends on the used PLC signal's sampling frequency, where a high value is recommended to offer the possibility of detecting all the received paths, especially for dense typologies with many potential and successive paths. In this case, it is likely that the CIR will get saturated throughout a path at each sample.

2. **VCIR Generation**
   Using steps 2 and 3 of Algorithm 1, the generation of VCIR and its multilevel quantization should be performed between each pair of endpoints in the considered topology to generate the corresponding keys.

3. **Reconciliation**
   The reconciliation process is carried out by this function using the position vector and key as inputs, as described in Sect. 7.2.3.
4. **Bit Mismatch Rate (BMR)**
   The BMR is used to assess the reciprocity of the proposed scheme when compared to that of the single-level quantization scheme. In particular, this metric can be defined as the ratio of the number of bits that do not match between the uplink and downlink IDs of a given communication link to the ID length.
5. **Correlation Evaluation**
   This evaluation consists of testing the independence between the different link IDs. Hence, the correlation coefficients between the generated ID of a given link and all other links are calculated. This metric presents a crucial condition for different PLS applications, including key generation and link identification.
6. **Secrecy Outage Probability (SOP)**
   The SOP metric can be used to determine how secure a legitimate communication is within a given adversary model. In this chapter, this adversary model can be generated for each topology by selecting two given nodes as legitimate transmitters and receivers and any other node as an eavesdropper. Accordingly, the SOP can be evaluated as the probability of the event when the corresponding Secrecy Capacity (SC) is less than a predefined target secrecy rate, denoted by SR_{Th}. The Secrecy Capacity (SC) is the difference between the instantaneous ergodic capacities of the legitimate channel (EC_L) and that of the eavesdropper channel (EC_{ϵ}). Therefore, the SOP can be expressed as follows:

$$SOP = Pr\left(\left[EC_L - EC_{\epsilon}\right] < SR_{\mathrm{Th}}\right). \tag{7.14}$$

The developed simulator has taken into account the imperfect channel conditions, where the CIR variations are estimated but not error-free. This is due to the fact that the channel probing is subject to Gaussian and impulsive noises, which may add or remove some impulses to affect the reciprocity of the quantization scheme. In view of this, the simulation process has been designed to conduct the aforementioned functions for a given number of iterations, N_{max}, for each given noise power value. Without loss of generality, the used simulation parameters' values are presented in Table 7.2.

7.3.2 Simulation Results

In this section, we discuss and assess the uniqueness of the generated keys, where we evaluate the correlation coefficients between all links under different typologies and noise levels.

Figures 7.3, 7.4, 7.5, 7.6, 7.7 and 7.8 present these simulation results that are structured as follows:

Table 7.2 The simulation parameter values [87]

Parameter	Value
N_{max}	10^6
The transmit power [dB]	126 dBμV
The average noise power [dBμV]	20 : 2 : 100
The sampling time: T_s	0.05 μs
The channel probing period	$256 \times T_s$
The quantization level: $L1$	0.02
The PL propagation attenuation factor: α [dB/m]	−30
The reflection attenuation factor: ρ [dB]	−20
The discontinuity attenuation factor: δ [dB]	−10
The maximum ID key length	128

1. In Figs. 7.3a, 7.5a, and 7.7a, the topology under consideration is illustrated, where the communicating nodes are presented in green and the cable discontinuities in red.
2. Without loss of generality, Figs. 7.3b, 7.5b, and 7.7b present the VCIR multilevel quantization and the relevant final key (after the reconciliation procedure) for the link ($A \leftrightarrow D$).
3. Figures 7.3c, 7.5c and 7.7c present the BMRs of the ($A \leftrightarrow D$) link versus the different considered noise level.
4. The variations of the correlation coefficients between the ($A \rightarrow D$) link ID and all other link IDs versus noise level are shown in Figs. 7.4, 7.6, and 7.8.

As presented in the simulation result's figures, the ($A \leftrightarrow D$) link's ID is highly positively correlated with the ($D \leftrightarrow A$) link's ID, which is not the case with the other link's IDs. This result is general for all the considered topologies, which confirm the uniqueness and the reciprocity of the generated PL link ID's.

In addition and for all the considered topologies, it is shown that the BMR is inversely proportional to the noise level, which is expected. This is because, when the noise power increases, the CIR estimation error increases, which decreases the VCIR and the final generated key accuracy.

A comparison of the BMR between the two quantization schemes reveals considerable disparities. In fact, for the noise power below 90 dBμV, the least BMR for the single-level quantization is higher than 10^{-3}. However, the proposed scheme achieves zero values for the same noise power range. This difference in BMR can be explained by the fact that the reconciliation phase of the multilevel quantization process reduces bit mismatches, which is not an option for one-level quantization. The inability to apply a reconciliation method has a consequence that the BMR of the one-level quantization will reach a minimum value that cannot be decreased, even if the noise levels are reduced.

For noise power greater than 90 dBμV, the BMR results are comparable for both quantization approaches, implying that the effect of the channel estimation error caused by high noise levels has an equivalent effect on both of them.

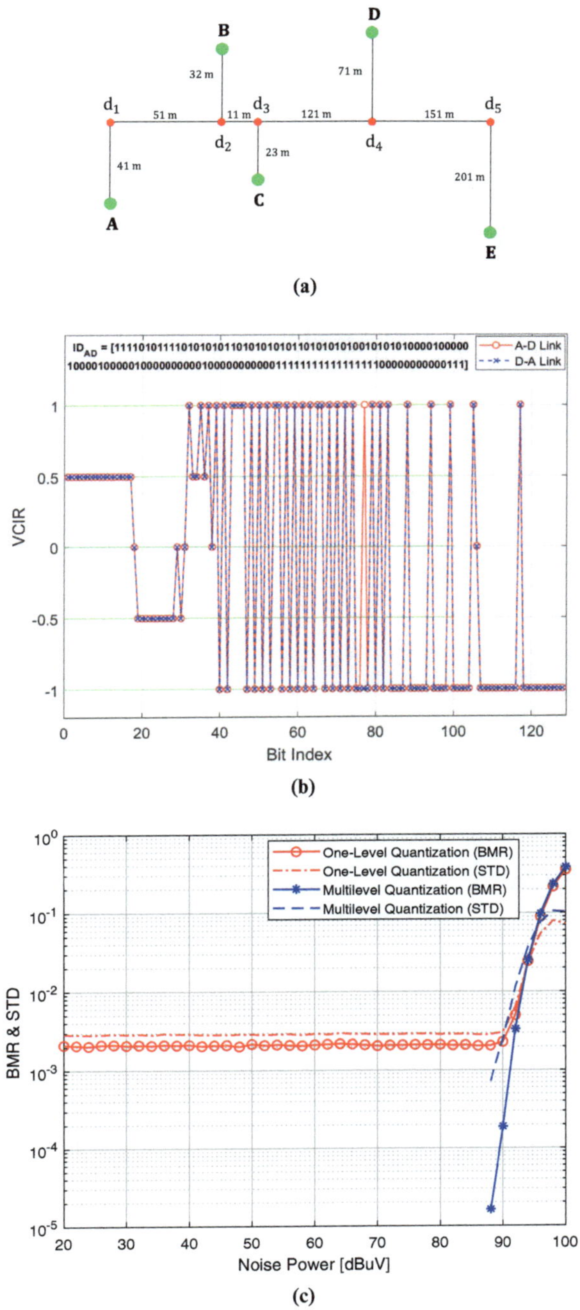

Fig. 7.3 Simulation results of a bus topology case: (**a**) the considered bus topology, (**b**) the VCIRs and IDs of $(A \leftrightarrow D)$ links, and (**c**) BMRs of the introduced multilevel quantization and the single-level quantization techniques [87]

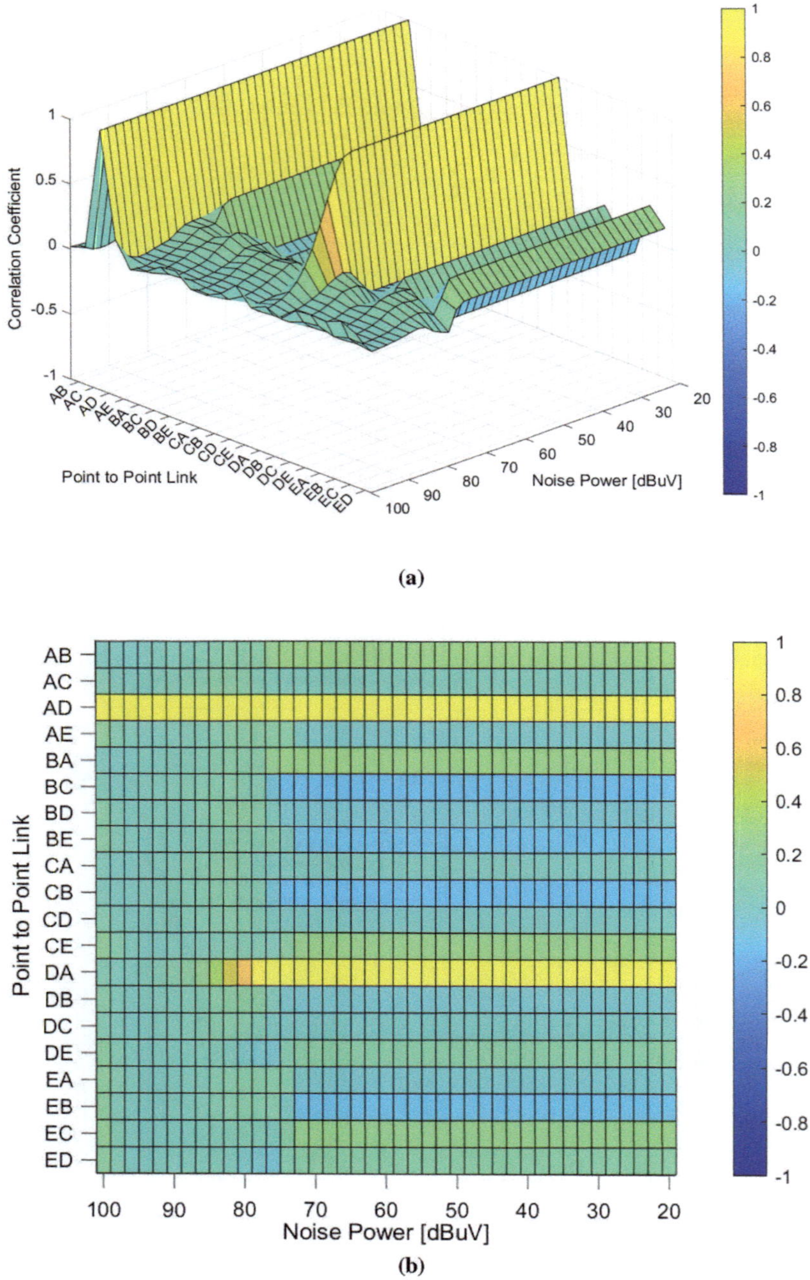

Fig. 7.4 The correlation coefficients between the $(A \rightarrow D)$ link's ID and the all links' IDs for the considered bus topology case [87]: (**a**) in a 3D space and (**b**) in a 2D space [87]

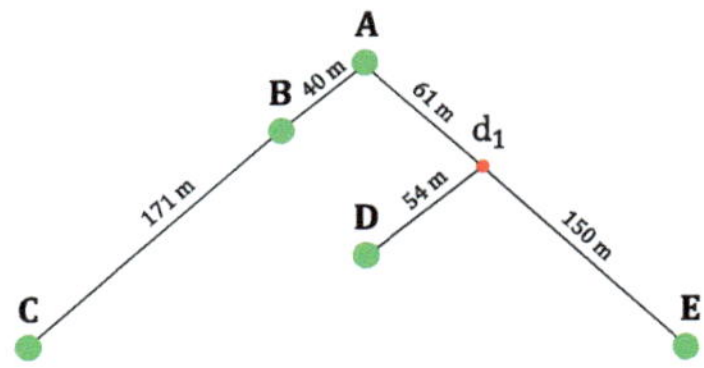

(a)

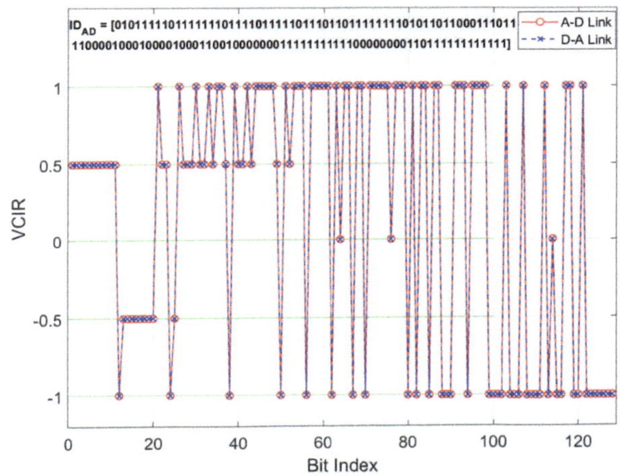

(b)

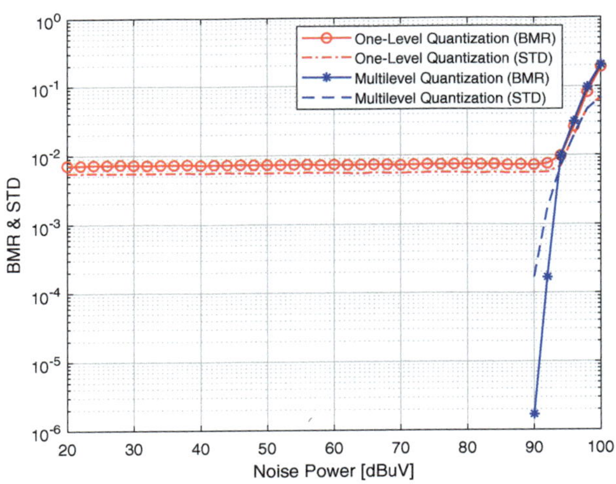

(c)

Fig. 7.5 Simulation results of a tree topology case: (**a**) the considered tree topology, (**b**) the VCIRs and IDs of $(A \leftrightarrow D)$ links, and (**c**) BMRs of the introduced multilevel quantization and the single-level quantization techniques [87]

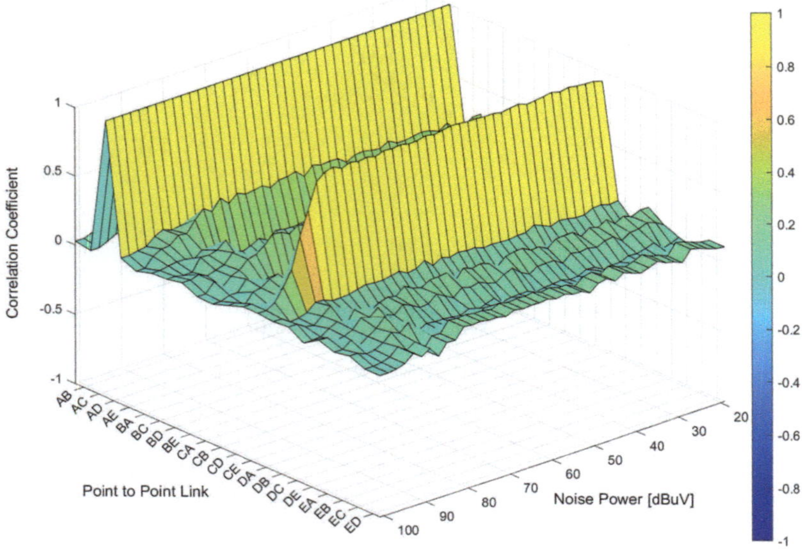

(a)

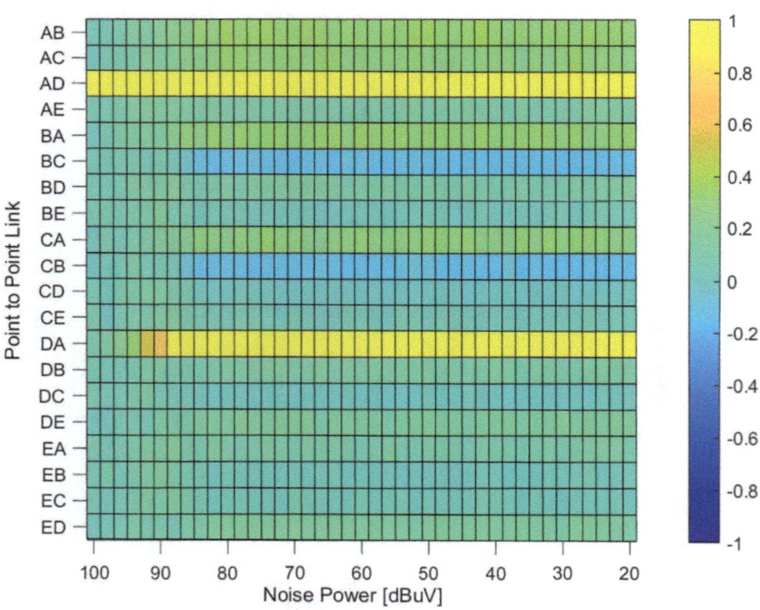

(b)

Fig. 7.6 The correlation coefficients between the $(A \rightarrow D)$ link's ID and the all links' IDs for the considered tree topology case [87]: (**a**) in a 3D space and (**b**) in a 2D space [87]

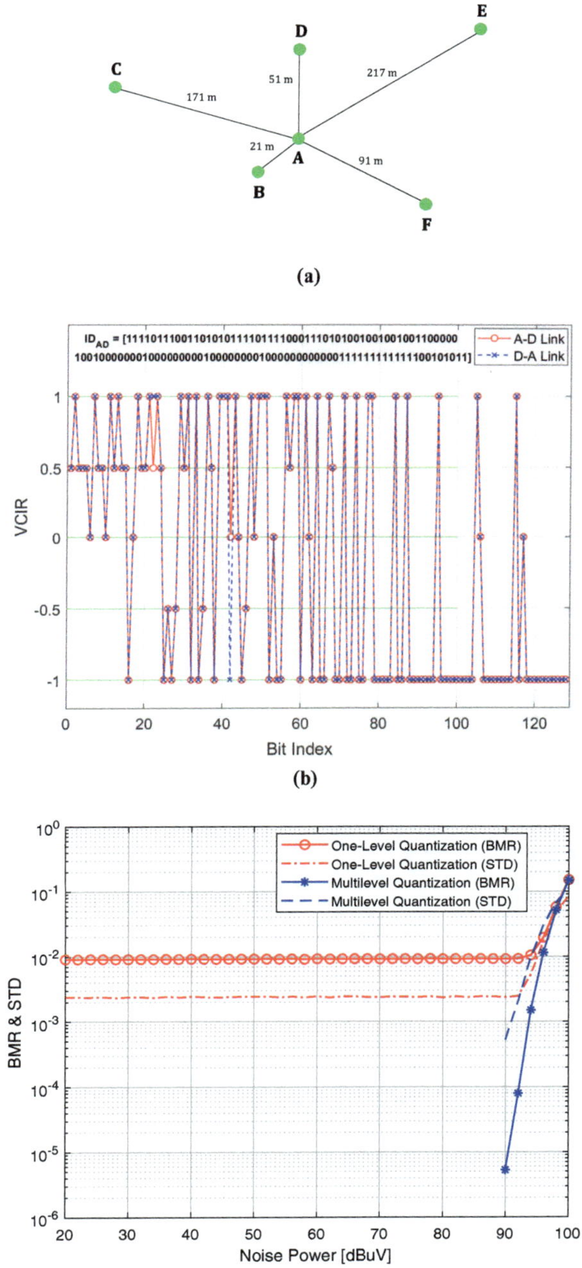

Fig. 7.7 Simulation results of a star topology case: (**a**) the considered star topology, (**b**) the VCIRs and IDs of $(A \leftrightarrow D)$ links, and (**c**) BMRs of the introduced multilevel quantization and the single-level quantization techniques [87]

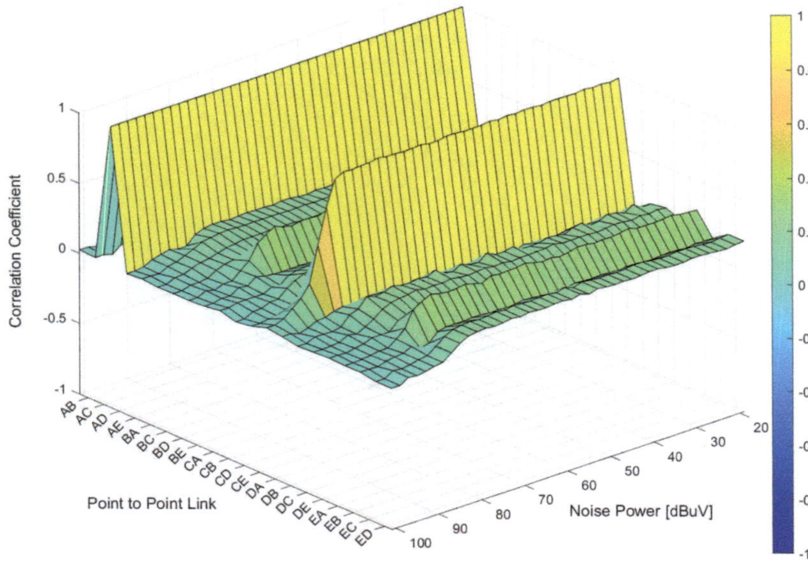

(a)

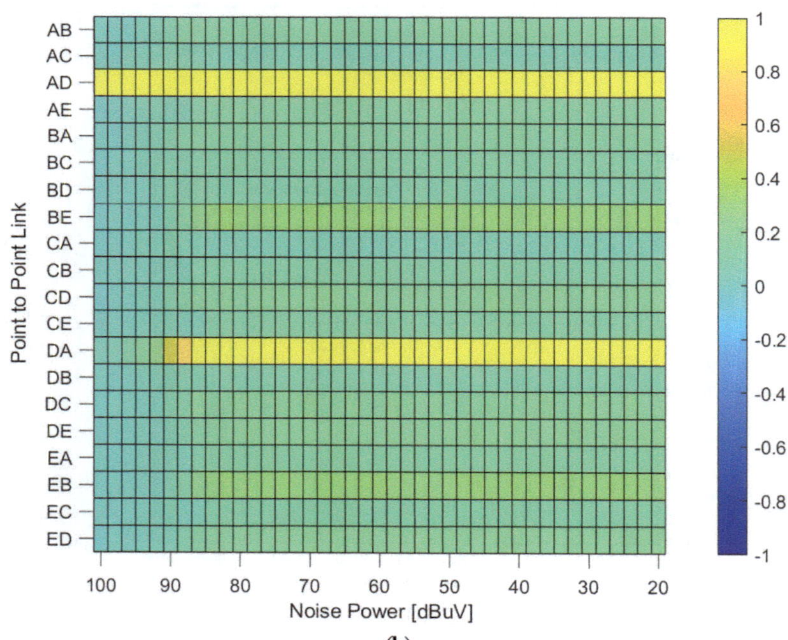

(b)

Fig. 7.8 The correlation coefficients between the $(A \to D)$ link's ID and the all links' IDs for the considered star topology case [87]: (**a**) in a 3D space and (**b**) in a 2D space [87]

The BMR and correlation coefficients for the considered topologies demonstrate that the output of the proposed technique can significantly enhance the security of legitimate communication in terms of SOP. Indeed, the simulation results show that the correlation between the keys of the considered legitimate nodes A and D is clear, which is not the case with the other link's keys, which can be used for eavesdropping attacks. Consequently, the secrecy rate can be easily increased, and hence, a decrease in the SOP can be observed.

In view of these simulation results and the clear impacts of the topology on the CIR, several research paths can be derived from this work's outcome, in particular, security amplification and topology inference techniques.

Chapter 8
Authentication Method for PLC Network

Abstract Authentication techniques in Power Line Communication (PLC) networks are vital to ensure secure communication, preventing unauthorized access and potential tampering with the data being transmitted. This chapter presents a novel technique based on multipath channel delays to offer Physical Layer Identification (PL ID) for PLC links. The method's precision and efficiency are tested by measuring the Successful Path Detection Probability (SPDP) within a simulated environment, accounting for both perfect and imperfect channel state information scenarios. The findings show that for PLC noise power of around 90 dBuV, the achieved accuracy is approximately 90%. When the noise power drops below 80 dBuV, accuracy peaks at 100%. The proposed PL ID technique serves as an auxiliary layer of authentication within PLC networks by providing a unique fingerprint for each link.

8.1 PLC Channel Impulse Response Model

This section presents the characteristics and modeling concepts of Power Line Communication (PLC)'s Channel Impulse Response (CIR) that will be instrumental for the remainder of this chapter to detail the principle of the proposed Physical Layer Identification (PL ID) technique.

To this end, the four scenarios presented in Fig. 8.1 have been considered to simulate the relevant impulse responses that have been used within the technique introduced in this chapter.

Let us begin by examining the initial situation shown in Fig. 8.1.a, the baseline scenario, in which a node A is connected directly to a node B without any discontinuity. For this scenario, the signal follows a direct path from A to B and an infinite number of secondary paths that result from the signal bouncing between A and B multiple times. Specifically, when $i = 0$, the direct path is $(A \rightarrow B)$. For the second path (first bounce, $i = 1$), it is $(A \rightarrow B \rightarrow A \rightarrow B)$, and so on.

J. Hernandez Fernandez et al., *Physical Layer Security in Power Line Communications*, Advances in Information Security 108,
https://doi.org/10.1007/978-3-031-57349-1_8

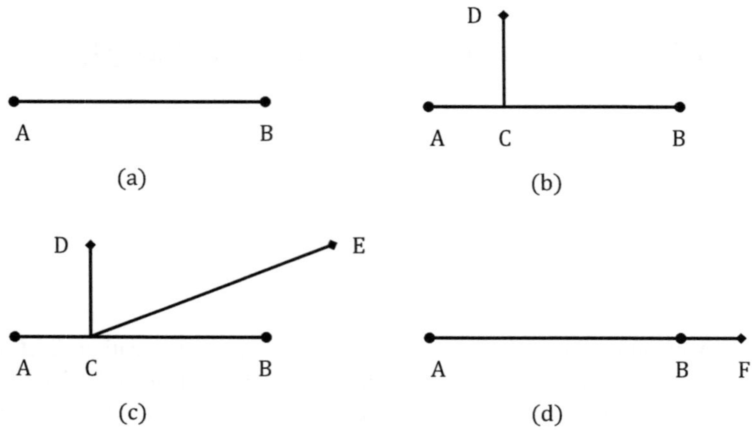

Fig. 8.1 PL system connection scenarios: (**a**) scenario 1, (**b**) scenario 2, (**c**) scenario 3, and (**d**) scenario 4 [88]

Let L_{XY} denote the distance between two nodes X and Y, and the propagation attenuation coefficient per power line length unit is represented by α, while the reflection and the discontinuity attenuation coefficients at node X are denoted by ρ_X and δ_X, respectively.

Accordingly, the relevant path lengths l_i and weights g_i for this scenario are presented in Table 8.1, where only the first 10 paths are considered, assuming that the remaining paths are significantly attenuated.

Consequently, a generic script has been developed to generate the delays and gains of the significant paths for each scenario. Accordingly, the power delay profiles of the four scenarios are presented in Table 8.1 and Figs. 8.2, 8.3, 8.4, and 8.5.

The presented characteristics and modeling of PLC CIR are the key to the introduced PL ID technique. Accordingly, the next section presents the proposed scheme in detail.

8.2 Physical Layer Identification Scheme Description

In this section, we provide a comprehensive description of the PL ID scheme. The proposed scheme is an accurate physical layer identification method for the fingerprinting of a given link between two nodes in a considered grid, using realistic (imperfect known of) CSIs. In particular, the proposed technique involves taking the mean of successive estimated CIRs over time to reduce the impact of noise [63, 204]. Indeed, it is well known that the PLC channel is slow-fading and thus remains relatively constant over time. In addition, by assuming that the channel estimation error follows a random distribution with a mean of zero, meaning that

Table 8.1 Power delay profiles of the ($N = 10$) first arrived paths for the different scenarios [88]

#	Path type	Path index	l_i	g_i
1	A-B-i(A-B)	$i \in [0, 10]$	$L_{AB}(1+2i)$	$(1-\alpha L_{AB})^{2i+1}(1-\rho_A)^i(1-\rho_B)^i$
2	A-C-B	$i=0$	L_{AB}	$(1-\alpha L_{AC})(1-\alpha L_{CB})(1-\delta_C)$
	A-C-A-C-B	$i=1$	$3L_{AC}+L_{CB}$	$(1-\alpha L_{AC})^3(1-\alpha L_{CB})(1-\rho_A)(1-\rho_C)(1-\delta_C)$
	A-C-D-C-B	$i=2$	$L_{AC}+2L_{CD}+L_{CB}$	$(1-\alpha L_{AC})(1-\alpha L_{CD})^2(1-\alpha L_{CB})(1-\rho_D)(1-\delta_C)^2$
	A-C-A-C-A-C-B	$i=3$	$5L_{AC}+L_{CB}$	$(1-\alpha L_{AC})^5(1-\alpha L_{CB})(1-\rho_A)^2(1-\rho_C)^2(1-\delta_C)$
	A-C-D-C-A-C-B	$i=4$	$3L_{AC}+2L_{CD}+L_{CB}$	$(1-\alpha L_{AC})^3(1-\alpha L_{CD})^2(1-\alpha L_{CB})(1-\rho_A)(1-\rho_D)(1-\delta_C)^3$
	A-C-A-C-D-C-B	$i=5$	$3L_{AC}+2L_{CD}+L_{CB}$	$(1-\alpha L_{AC})^3(1-\alpha L_{CD})^2(1-\alpha L_{CB})(1-\rho_A)(1-\rho_C)(1-\rho_D)(1-\delta_C)^2$
	A-C-B-C-B	$i=6$	$L_{AC}+3L_{CB}$	$(1-\alpha L_{AC})(1-\alpha L_{CB})^3(1-\rho_B)(1-\rho_C)(1-\delta_C)$
	A-C-D-C-D-C-B	$i=7$	$L_{AC}+4L_{CD}+L_{CB}$	$(1-\alpha L_{AC})(1-\alpha L_{CD})^4(1-\alpha L_{CB})(1-\rho_C)(1-\rho_D)^2(1-\delta_C)^2$
	A-C-A-C-B-C-B	$i=8$	$3L_{AC}+3L_{CB}$	$(1-\alpha L_{AC})^3(1-\alpha L_{CB})^3(1-\rho_A)(1-\rho_B)(1-\rho_C)^2(1-\delta_C)$
	A-C-B-C-A-C-B	$i=9$	$3L_{AC}+3L_{CB}$	$(1-\alpha L_{AC})^3(1-\alpha L_{CB})^3(1-\rho_A)(1-\rho_B)(1-\delta_C)^3$

(continued)

Table 8.1 (continued)

#	Path type	Path index	l_i	g_i
3	A-C-B	$i=0$	L_{AB}	$(1-\alpha L_{AC})(1-\alpha L_{CB})(1-\delta_C)$
	A-C-A-C-B	$i=1$	$3L_{AC}+L_{CB}$	$(1-\alpha L_{AC})^3(1-\alpha L_{CB})(1-\rho_A)(1-\rho_C)(1-\delta_C)$
	A-C-D-C-B	$i=2$	$L_{AC}+2L_{CD}+L_{CB}$	$(1-\alpha L_{AC})(1-\alpha L_{CD})^2(1-\alpha L_{CB})(1-\rho_D)(1-\delta_C)^2$
	A-C-A-C-A-C-B	$i=3$	$5L_{AC}+L_{CB}$	$(1-\alpha L_{AC})^5(1-\alpha L_{CB})(1-\rho_A)^2(1-\rho_C)^2(1-\delta_C)$
	A-C-D-C-A-C-B	$i=4$	$3L_{AC}+2L_{CD}+L_{CB}$	$(1-\alpha L_{AC})^3(1-\alpha L_{CD})^2(1-\alpha L_{CB})(1-\rho_A)(1-\rho_D)(1-\delta_C)^3$
	A-C-A-C-D-C-B	$i=5$	$3L_{AC}+2L_{CD}+L_{CB}$	$(1-\alpha L_{AC})^3(1-\alpha L_{CD})^2(1-\alpha L_{CB})(1-\rho_A)(1-\rho_C)(1-\rho_D)(1-\delta_C)^2$
	A-C-B-C-B	$i=6$	$L_{AC}+3L_{CB}$	$(1-\alpha L_{AC})(1-\alpha L_{CB})^3(1-\rho_B)(1-\rho_C)(1-\delta_C)$
	A-C-D-C-D-C-B	$i=7$	$L_{AC}+4L_{CD}+L_{CB}$	$(1-\alpha L_{AC})(1-\alpha L_{CD})^4(1-\alpha L_{CB})(1-\rho_C)(1-\rho_D)^2(1-\delta_C)^2$
	A-C-A-C-E-C-B	$i=8$	$3L_{AC}+2L_{CE}+L_{CB}$	$(1-\alpha L_{AC})^3(1-\alpha L_{CE})^2(1-\alpha L_{CB})(1-\rho_A)(1-\rho_C)(1-\rho_E)(1-\delta_C)^2$
	A-C-E-C-E-C-B	$i=9$	$L_{AC}+4L_{CE}+L_{CB}$	$(1-\alpha L_{AC})(1-\alpha L_{CE})^4(1-\alpha L_{CB})(1-\rho_C)(1-\rho_E)^2(1-\delta_C)^2$
4	A-B	$i=0$	L_{AB}	$(1-\alpha L_{AB})$
	A-B-F-B	$i=1$	$L_{AB}+2L_{BF}$	$(1-\alpha L_{AB})(1-\alpha L_{BF})^2(1-\rho_F)(1-\delta_B)$
	A-B-F-B-F-B	$i=2$	$L_{AB}+4L_{BF}$	$(1-\alpha L_{AB})(1-\alpha L_{BF})^4(1-\rho_B)(1-\rho_F)^2(1-\delta_B)$
	A-B-F-B-F-B-F-B	$i=3$	$L_{AB}+6.*L_{AB}$	$(1-\alpha L_{AB})(1-\alpha L_{BF})^6(1-\rho_B)^2(1-\rho_F)^3(1-\delta_B)$
	A-B-A-B	$i=4$	$3L_{AB}$	$(1-\alpha L_{AB})^3(1-\rho_A)(1-\rho_B)$
	A-B-A-B-F-B	$i=5$	$3L_{AB}+2L_{BF}$	$(1-\alpha L_{AB})^3(1-\alpha L_{BF})^2(1-\rho_A)(1-\rho_B)(1-\rho_F)(1-\delta_B)$
	A-B-F-B-A-B	$i=6$	$3L_{AB}+2L_{BF}$	$(1-\alpha L_{AB})^3(1-\alpha L_{BF})^2(1-\rho_A)(1-\rho_F)^2(1-\delta_B)^2$
	A-B-F-B-A-B-F-B	$i=7$	$3L_{AB}+4L_{BF}$	$(1-\alpha L_{AB})^3(1-\alpha L_{BF})^4(1-\rho_A)(1-\rho_F)^2(1-\delta_B)^3$
	A-B-A-B-F-B-F-B	$i=8$	$3L_{AB}+4L_{BF}$	$(1-\alpha L_{AB})^3(1-\alpha L_{BF})^4(1-\rho_A)(1-\rho_B)^2(1-\rho_F)^2(1-\delta_B)$
	A-B-A-B-A-B	$i=9$	$5L_{AB}$	$(1-\alpha L_{AB})^5(1-\rho_A)^2(1-\rho_B)^2$

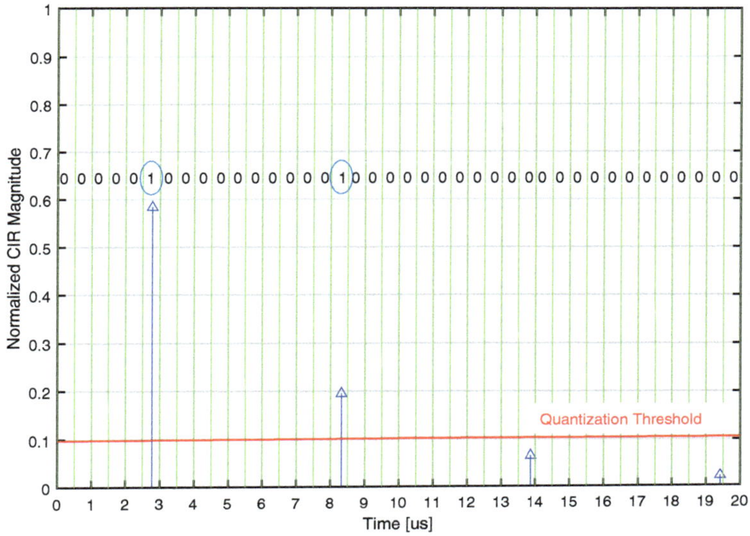

Fig. 8.2 CIR of scenario 1 [88]

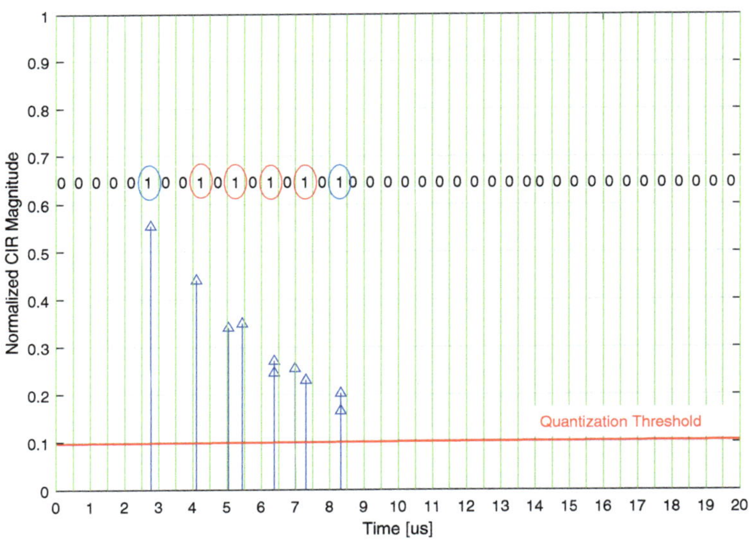

Fig. 8.3 CIR of scenario 2 [88]

this error can have either a negative or a positive value, the corresponding sum will be close to zero. Consequently, the average of the different CIRs can be used to reduce the channel estimation errors.

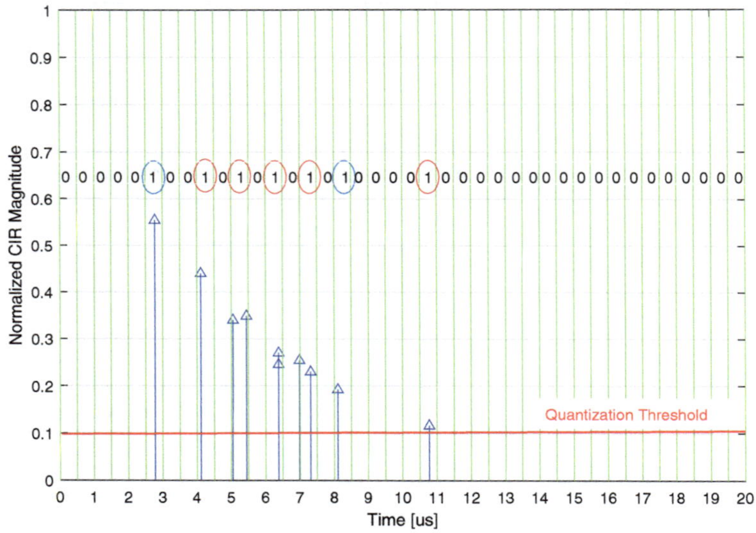

Fig. 8.4 CIR of scenario 3 [88]

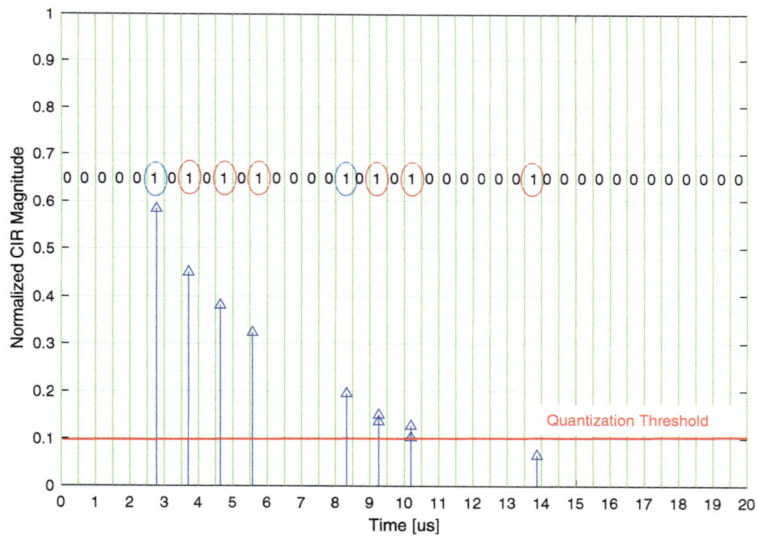

Fig. 8.5 CIR of scenario 4 [88]

To further detail the proposed scheme, we present in Algorithm 2 the following main relevant steps:

- **Step 1:** This step consists of a channel sensing process that identifies the corresponding CIR of the considered link between two given nodes. Specifically,

Algorithm 2 Physical Layer Identification (PL ID)

Inputs: PL ID Length: ID_{Len}
Sampling Time: T_S
Quantization Threshold
Number of observations: N_{Obs}
Received Signals: $y(n)$, $n \in \{1, ..N_{Obs}\}$
Step 1: Channel Probing
Estimation of the different N_{Obs} CIRs: $\hat{h}(n)$
Step 2: Minimizing the Channel Estimation Error:

$$\tilde{h} = \frac{\sum_{n=1}^{N_{Obs}} \hat{h}(n)}{N_{Obs}}$$

Step 3: CIR quantization to generate the PL ID.

by using N_{Obs} received signals, the corresponding CIRs $\hat{h}(n)$, $n \in \{1, ..N_{Obs}\}$, can be estimated. In this work, the node access control is not addressed, where it is assumed that all legitimate nodes in the considered network are registered and synchronized in order to accurately estimate the CIRs.

- **Step 2:** In this step, the channel estimation error can be minimized by taking the average of the N_{Obs} estimated CIRs. This step is essential in order to provide an accurate PL ID and to increase the received Signal-to-Noise Ratio (SNR), thus enhancing the overall quality of data transmission.
- **Step 3:** Finally, a standard quantization should be provided in this step to generate the corresponding PL ID.

Figures 8.2, 8.3, 8.4, and 8.5 show the corresponding normalized CIR magnitudes for the four scenarios previously presented.

8.3 Simulation Results and Discussion

In this section, the numerical results and advantages of the proposed PL ID technique are presented and evaluated. To this end and without loss of generality, scenario 4 of Fig. 8.1 has been considered, along with the following parameters: $L_{AB} = 416$, $L_{AC} = 100$ m, $L_{CB} = 316$ m, $L_{CD} = 170$ m, $L_{CE} = 300$ m, $L_{BF} = 70$ m, $\alpha = -30$ dB/m, $\rho_X = -20$ dB, $\delta_X = -10$ dB, $T_S = 0.5$ μs, the PL ID length $= 32$, and the average PLC noise power range from 50 dBμV to 110 dBμV [63, 177, 204].

The normalized CIR for the introduced technique, along with the perfect and imperfect normalized CIRs, is presented in Fig. 8.6. As shown in this figure, the generated CIR by the proposed scheme is almost the same as that of the corresponding perfect CIR, which confirms its advantages in terms of accurate CIR estimation and accurate PL ID.

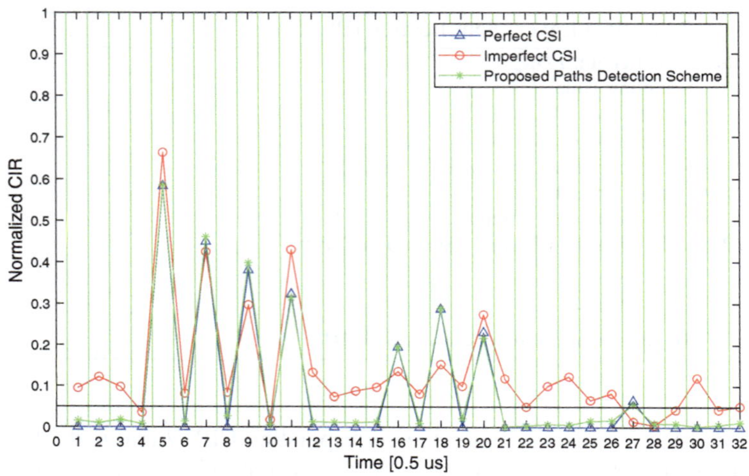

Fig. 8.6 The CIR variations versus time for the introduced method, as well as for the perfect and imperfect CSI scenarios [88]

To evaluate the advantages of the introduced scheme, when compared to related work in the literature, we present in Fig. 8.7 a comparison of the Successful Path Detection Probability (SPDP) between the introduced scheme and one of the most relevant techniques presented in [186]. We will refer to the cited technique as PT scheme, which is derived from the surnames of the authors: Passerini and Tonello. In this comparison, the SPDP is defined as the probability of the event when the number of detected paths is greater than or equal to a predefined path detection threshold (PD_{th}).

As shown in Fig. 8.7, the SPDP of the introduced technique is larger than that of the PT technique for all the different values of PD_{th}, which confirms the advantages of the introduced technique in terms of path detection accuracy. In particular, the results confirm the robustness and accuracy of the introduced technique, even for noisy PLC environments, which is not the case for the PT technique. In fact, as shown in this figure, an average gain (in terms of noise power) of more than 15 dB can be reached by using the proposed scheme to offer the same SPDP as the PT scheme. As an example, for as SPDP value equal to 0.9 and a given $PD_{\text{th}} = 90\%$, the PT scheme can provide the desired SPDP threshold, when the noise power is up to 71 dBμV. However, the proposed scheme can achieve the same target when the noise power is up to 89 dBμV, with a noise power gain of 18 dB. Moreover, it is shown that the SPDP is inversely proportional to the values of PD_{th}, which is expected.

Figure 8.8 presents the variations of the Successful Path Detection Gain (SPDG) versus noise power for the introduced and the PT schemes, with different values of transmission delay. In this work, the SPDG is defined as the difference between the proposed scheme's SPDP and that of the PT scheme in percentage units.

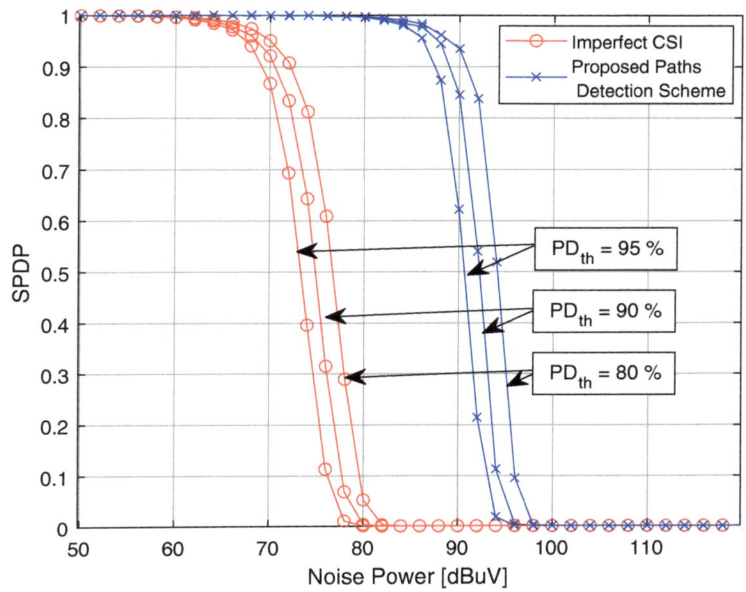

Fig. 8.7 SPDP variations versus noise power, for the introduced and the PT schemes, with $N_{\text{obs}} = 100$, and for different values of PD_{th} [88]

The mentioned transmission delay in this graph is related to the used Orthogonal Frequency Division Multiplexing (OFDM) symbol duration and the number of observations N_{Obs}. Without loss of generality, the PRIME technology's OFDM symbol duration is considered, which is equal to 2.048 ms [201].

According to the proposed scheme, the same number of OFDM symbols as the number of observations (N_{Obs}) is needed to estimate and conduct the average of N_{Obs} CIRs. This results in a delay of N_{Obs} times the duration of the OFDM symbol. For these simulation results, different observation values were used, such as $N_{Obs} = \{2,\ 5,\ 10,\ 20,\ 50,\ 100\}$, which correspond to transmission delays of 4.09, 10.24, 20.48, 40.96, 102.4, and 204.8 ms, respectively.

As shown in Fig. 8.8, for the various observation numbers/transmission delays, when the noise power increases, the SPDG rises until it reaches a certain maximum, and then it decreases and converges to zero.

To explain this SPDG variation, let the range of noise power between NP_{min} and NP_{max} present the efficient noise power range, where the introduced scheme offers better performance than the PT scheme in terms of SPDP. For a noise power level below NP_{min}, the impact of noise on the accuracy of channel estimation is insignificant for both approaches, i.e., the SPDP has reached the value of 1 in both cases, as demonstrated in Fig. 8.7. In this case, no gain in terms of successful path detection can be observed. Therefore, the SPDG is zero.

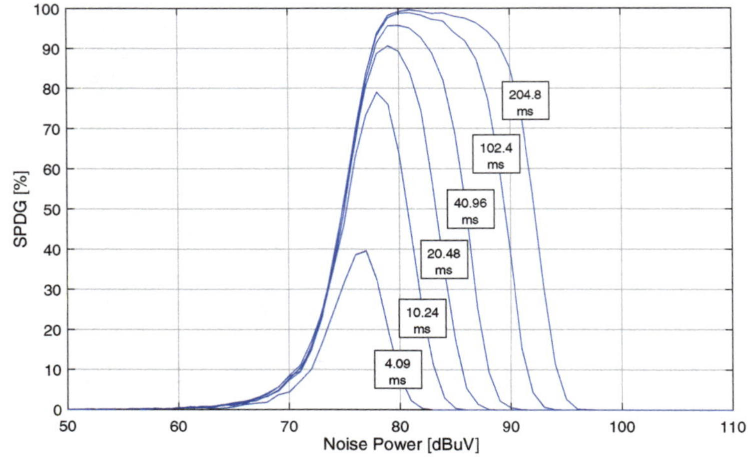

Fig. 8.8 The SPDG variation of the introduced technique versus transmission delay, with respect to the PT scheme [88]

By increasing the noise power to values greater than NP_{min}, the proposed technique outperforms the PT scheme, resulting in an increase of the relevant SPDG. This is because the introduced technique offers a better channel estimation error when compared to the PT scheme in this case.

As the noise power increases, the advantage of our scheme compared to the PT scheme gradually diminishes, eventually reaching zero. This is because a higher noise power has a significant impact on the accuracy of both schemes in terms of SPDP, which is zero for both schemes when the noise power is greater than or equal to NP_{max}. In addition, the graph shows that the efficient noise power range and its relevant maximum SPDG increase as the number of observations increases, which is expected. This can be explained by the fact that a large number of observations improve the accuracy of the estimated CIR, which results in a significant gain with respect to the PT scheme. In contrast, increasing the number of observations leads to a transmission delay. Nevertheless, this can be disregarded since the potential applications of the proposed scheme usually do not require a real-time response. Furthermore, the cited delays are in the order of milliseconds. For example, as shown in this figure, for a transmission delay of only 204.8 ms, a noise power efficient range of 30 dB with a successful path detection gain of more than 90 % can be observed.

The variation of the Bit Mismatch Rate (BMR) versus the noise power is presented in Fig. 8.9, with different numbers of observations, e.g., 2, 5, 10, 20, 50, and 100. The BMR is defined as the ratio of mismatched bits between the PL IDs of the proposed and the PT schemes over the PL ID length. As depicted in this figure,

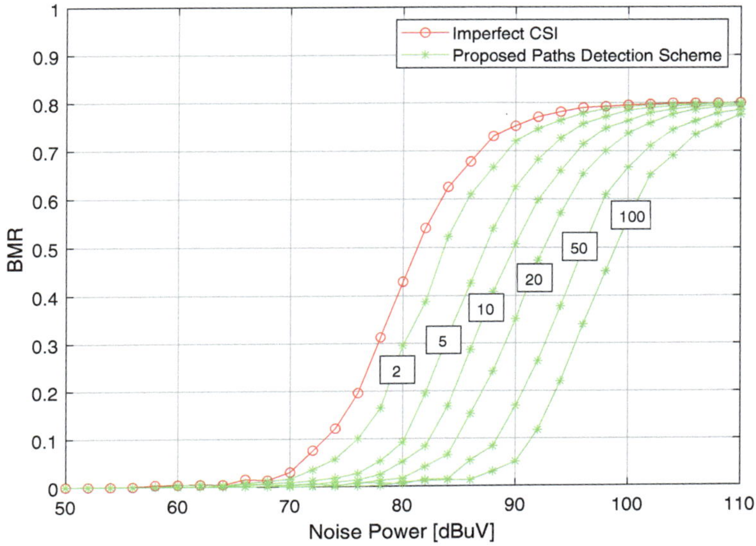

Fig. 8.9 The BMR variations of the introduced and the PT schemes, with the different numbers of observations: 2, 5, 10, 20, 50, and 100 [88]

increasing the number of observations results in an increase of the BMR. This is due to the fact that the channel estimation error is inversely proportional to the number of observations, as previously explained.

Chapter 9
PLC Network Integrity Solution

Abstract Assuring that the power line has not been altered or tampered with in an unauthorized manner is key for maintaining the integrity of Power Line Communication (PLC) networks. This chapter introduces a PLC topology change detection technique that can identify modifications in the power grid physical network. Experimental results are presented to confirm the quality and viability of the proposed approach, showing excellent detection accuracy, even at high PLC noise levels. This Channel Impulse Response (CIR)-only-based solution for topology detection presents an efficient PLC network integrity scheme that cannot only detect but also identify the specific altered links.

9.1 Background on PLC Topology Change Detection Solutions

The concept of smart grids involves the integration of advanced technology into electrical grids to optimize performance across various domains, such as generation, transmission, distribution, and consumer applications. In this context, Power Line Communication (PLC) has emerged as a promising telecommunication technology for modernizing the power grid. In fact, by reusing existing network infrastructures, PLC offers a cost-efficient two-way communication between utility assets. This paves the way for developing a variety of applications, including topology estimations [140], grid management [35, 233], and energy loss detection [41, 42], etc.

However, due to their geographical dispersion and the continuous transformation brought on by automation and digitization, PLC-based smart grids are becoming more vulnerable to attacks [103]. In particular, the security of PLC-based smart grids presents a serious issue that has been of great interest to both the industry and the research community in recent years [250]. Consequently, the use of PLC security-based approaches is becoming increasingly widespread as a way to protect smart grids from malicious cyber-physical attacks, fraud, or accidental wrongdoings [153, 169].

J. Hernandez Fernandez et al., *Physical Layer Security in Power Line Communications*, Advances in Information Security 108,
https://doi.org/10.1007/978-3-031-57349-1_9

In view of this, utility companies can highly benefit from services that monitor and protect their assets, especially those related to the topology, as they meet both operational and security requirements. Indeed, the network's topology can be altered due to a variety of occurrences, some of which are legitimate, such as when linemen carry out network expansion projects or switch lines and others that are illegal, like unauthorized connections (nontechnical losses).

The main idea behind topology solutions consists of identifying the relevant network map that presents the topology physical links of deployed assets. These maps can be used to keep track of any equipment that is added or removed from the network, to detect anomalies, and to identify malicious intrusions while also optimizing the dispatch of crews to solve the issues. Given its natural integration with existing electrical infrastructure, PLC holds a distinctive advantage over competing technologies in their deployment and in their ability to ensure that the network map is accurate and up to date without the need for manual mapping of the network. Therefore, the required time and cost, as well as the efforts related to configuration and topology change detection, can be significantly reduced. Moreover, automated topology solutions are more reliable than manual techniques when it comes to security-related applications, as they minimize the possibility of human errors.

In addition to security enhancements, the technical and economic advantages of topology-based solutions have been demonstrated to be beneficial in preserving other grid components such as transformers [16, 90].

The possibility of reducing maintenance costs, increasing network reliability, performing robust network mapping, and providing real-time security features makes these applications a promising enhancement and solution for PLC-based smart grids.

In the literature, many relevant works have been introduced and evaluated in different PLC environments. Consequently, a variety of smart grid applications have been developed, taking advantage of the inherent information obtained from the PLC telecommunication signals, including faults and/or loss detection in the electrical grid [41, 42, 164, 184, 260].

These applications are classified into two categories: time domain-based and frequency domain-based techniques. The first category consists of analyzing the PLC signal in the time domain to extract the considered PLC channel time domain characteristics, e.g., the Channel Impulse Response (CIR). For the frequency domain-based technique, the PLC signal spectrum and the corresponding Channel Frequency Response (CFR) are analyzed to reveal the frequency domain characteristics of the considered PLC channels.

Further details about these PLC-based topology-related categories are highlighted in the following section, which presents the related works in the literature.

9.1.1 Time Domain-Based PLC Topology Solutions

The time domain-based approach relies on extracting PLC signal characteristics to identify the relevant network's topology. In particular, it consists of tracking and analyzing the PLC waveforms at different points in the network to estimate the PLC signal propagation time and hence identify the corresponding network's topology.

In this context, there are two main methods for time domain-based PLC topology recognition: the Time of Arrival (ToA) and the Time Difference of Arrival (TDoA). The main idea of the ToA method is to estimate the PLC signal propagation time between two known points of a given topology to identify the corresponding transmission distance (the PL length between the two points).

The TDoA can be considered as a positioning methodology that determines the difference between the time of arrival (ToA) of the PLC signal at two PLC devices located at the end of a given PLC branch. This approach can be used to accurately calculate the position of the connected devices and the relevant PL branch that connects them. These two approaches are commonly used in topology estimation solutions [14, 58, 219, 220], especially for inferring distances [57].

9.1.2 Frequency Domain-Based PLC Topology Solutions

The frequency domain-based methods are also further divided into two main categories: PLC signal spectral analysis and autocorrelation.

The first category involves the use of power spectral analysis to discern patterns, often in conjunction with Channel State Information (CSI) or Frequency Domain Reflectometry (FDR) with its two variants: the single-point FDR technique [4, 5] and multipoint FDR technique [6, 129, 185, 257].

Autocorrelation is employed to quantify the temporal correlation of PLC signals. The subsequent characterization of these PLC signals involves categorizing frequencies into distinct clusters or classes. In this context, the authors of [102] have proposed the use of time-series prediction to estimate the CSI, without the use of reference signal or the training phases required by machine learning approaches. In the study cited as [252], the authors explore the correlation between the CFR and the characteristics of the power line as a method to supervise the quality of electrical power.

Some works have combined both domains to introduce new efficient solutions [6, 129, 185, 257]. Other works have used electrical network parameters, including the admittance [183] or smart meter data [140, 188] to detect and identify topology changes. Different approaches have been presented in [99, 101, 195], where the authors have used neural networks and machine learning algorithms in combination with PLC channel state data to introduce new cable diagnosis and security applications. A noteworthy study is presented in [80], which shifts the focus from the physical cable topology to the communication topologies generated by PLC modems.

9.1.3 Comparison with Alternative Solutions

In this chapter, the path delays of the CIR are used to introduce a distributed and deterministic algorithm that can identify and locate changes in the power grid network. In the literature, a variety of PLC signal-based solutions are based on the CIR, such as a power cable diagnosis method [100], key generation technique [186, 195], and machine-learning-based PLC intrusion detection system [195].

The solution presented below shares the aforementioned technique, but it addresses the problem from a distributed and multipoint perspective without involving machine learning techniques. In particular, the proposed technique differs from the abovementioned related works, offering a fully CIR-based, distributed, and deterministic algorithm for the detection and localization of topological changes in power lines.

Before detailing the proposed technique, we present in the following section the considered PLC system and channel models.

9.2 PLC System and Channel Models

In this section, the considered PLC system and channel models are presented and detailed, where we focus on the generation of the relevant power delay profile that presents the PLC channel key feature for the introduced topology change detection approach.

9.2.1 PLC System Models

Two PLC system models are used to explain and evaluate the proposed topology change detection approach. The first model is a simple two-node topology commonly considered in the literature to present similar concepts. The second system model involves a more general multi-node topology.

9.2.2 Two-Node Topology Model

The two-node topology model consists of a direct connection between two given nodes, N_1 and N_2, with a middle branch at a given point X with a termination point Y, as shown in Fig. 9.1.

Fig. 9.1 Two-node topology model [89]

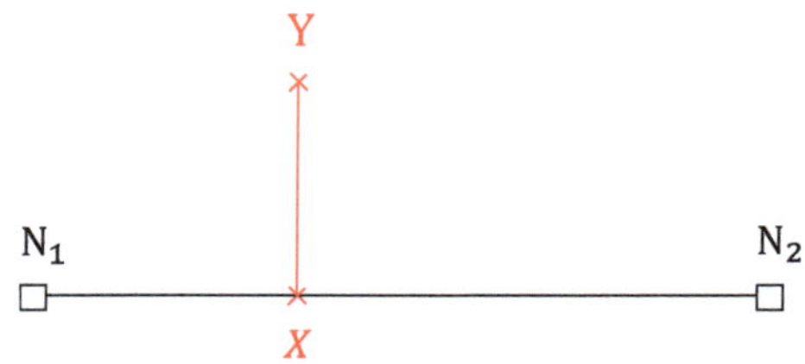

9.2.3 *Multi-Node Topology Model*

As shown in Fig. 9.2, the multi-node topology model includes a main backbone, where many nodes (N_i) are connected at various discontinuity points (d_i). Furthermore, two unknown nodes Y_1 and Y_2 are connected to the main power line at locations X_1 and X_2, respectively, where detecting these two unknown connections presents the main challenge of the technique introduced in this chapter.

9.3 PLC Multipath Channel Model

In this subsection, we detail the PLC multipath channel model for the two considered PLC networks. The time domain modeling approach [130] is used to generate the corresponding path delays and gains for each model/scenario.

Let us start by looking into the initial baseline situation shown in Fig. 9.1 without the link $(X - Y)$. In this case, the signal takes a first direct path $(N_1 \rightarrow N_2)$ and a set of significant secondary paths as a result of the signal bouncing i times between N_1 and N_2. More specifically, the first path $(N_1 \rightarrow N_2)$ presents the direct path index $i = 0$. Then, the secondary paths can be presented as $(N_1 \rightarrow N_2 \rightarrow i\ (N_1 \rightarrow N_2))$, for $i > 0$. Accordingly, Table 9.1 presents the significant path lengths l_i and weights g_i of each path index i, where $L_{N_1N_2}$ denotes the distance between nodes N_1 and N_2, α denotes the propagation attenuation coefficient per power line length unit, ρ_X denotes the reflection attenuation coefficient at node X, and δ_X denotes the discontinuity attenuation coefficient at node X.

Consequently, the power delay profile (lengths l_i and weights g_i) of the baseline scenario can be generated using the following expressions:

$$l_i = L_{N_1N_2}(1 + 2i) \tag{9.1}$$

and

$$g_i = (1 - \alpha L_{N_1N_2})^{2i+1}(1 - \rho_{N_1})^{i}(1 - \rho_{N_2})^{i}, \tag{9.2}$$

respectively.

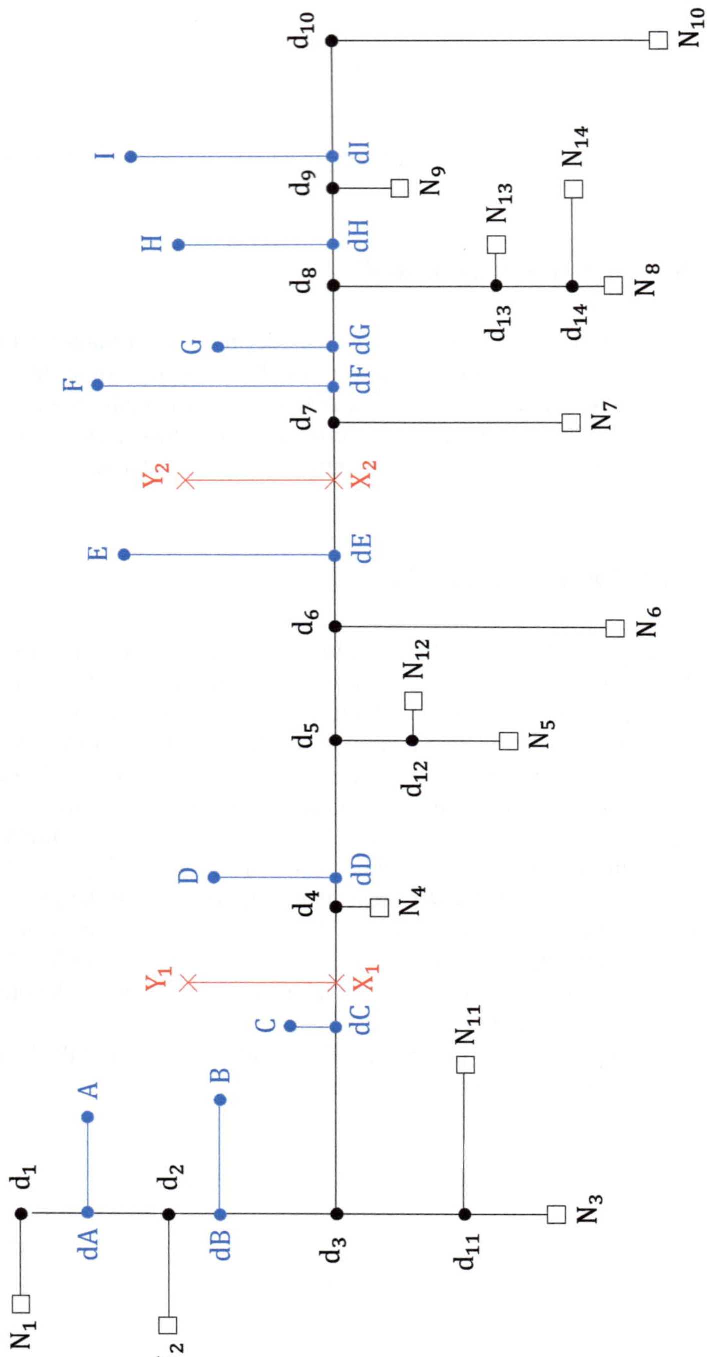

Fig. 9.2 An instance of a multi-node topology model [89]

Table 9.1 Power delay profile for the two-node topology model [89]

Path	Path index	l_i	g_i
$N_1 \rightarrow N_2$	$i = 0$	L_{N1N2}	$(1 - \alpha L_{N1N2})$
$N_1 \rightarrow N_2 \rightarrow N_1 \rightarrow N_2$	$i = 1$	$3L_{N1N2}$	$(1 - \alpha L_{N1N2})^3 (1 - \rho_{N_1}) (1 - \rho_{N_2})$
$N_1 \rightarrow N_2 \rightarrow i(N_1 \rightarrow N_2)$	i	$(2i + 1)L_{N1N2}$	$(1 - \alpha L_{N1N2})^{2i+1} (1 - \rho_{N_1})^i (1 - \rho_{N_2})^i$

Now, let us move on to the scenario depicted in Fig. 9.1, where the connection $(X - Y)$ is added. Following the same multipath channel modeling approach, an endless number of secondary paths arriving at N_2 can be generated, where Table 9.2 lists the relevant first ten significant path lengths l_i and weights g_i for this scenario.

In view of this, a general and flexible script has been developed and deployed to exhaustively generate the significant paths and the corresponding power delay profile between any given two nodes of the presented multi-node topology in Fig. 9.2.

9.4 Algorithmic Overview of the Topology Change Detection System

The methodology for detecting changes in network topology leverages the Physical Layer Identification (PL ID) previously discussed in this book. Specifically, it defines the procedure for generating a PL ID between two distinct nodes, N_i and N_j^i. In the context of the Topology Change Detection, the PL ID serves as a unique descriptor facilitating the identification of connections between nodes. In the forthcoming discussion, an introductory outline of the PL ID will be provided prior to an exhaustive exploration of the Topology Change Detection algorithm. For a more comprehensive treatment of PL ID and its critical role in the algorithm, readers are advised to consult Chap. 8. To facilitate a clear understanding of the algorithms discussed in this chapter, Table 9.3 enumerates the primary symbols utilized throughout the text.

9.4.1 Physical Layer Identification Algorithm

The inputs for Algorithm 3 are the PL ID length (ID_{Len}), the sampling time (T_S), the quantization threshold (Q_{th}), the number of observations (N_{Obs}), and the received signals ($y_{N_i,N_j^i}(n)$, $n \in \{1, ..N_{Obs}\}$).

Using these inputs, Algorithm 3 generates the PL identification of the link between N_i and N_j^i, by performing the following three main steps:

Table 9.2 Power delay profile of the $(N = 10)$ first arrived paths for the T-topology model [89]

Path	Path index	l_i	g_i
$N_1 - X - N_2$	$i = 0$	L_{N1N2}	$(1 - \alpha L_{N1X})\,(1 - \alpha L_{XN2})\,(1 - \delta_X)$
$N_1 - X - N_1 - X - N_2$	$i = 1$	$3L_{N1X} + L_{XN2}$	$(1 - \alpha L_{N1X})^3 (1 - \alpha L_{XN2})\,(1 - \rho_{N_1})\,(1 - \rho_X)\,(1 - \delta_X)$
$N_1 - X - Y - X - N_2$	$i = 2$	$L_{N1X} + 2L_{XY} + L_{XN2}$	$(1 - \alpha L_{N1X})\,(1 - \alpha L_{XY})^2 (1 - \alpha L_{XN2})\,(1 - \rho_Y)\,(1 - \delta_X)^2$
$N_1 - X - N_1 - X - N_1 - X - N_2$	$i = 3$	$5L_{N1X} + L_{XN2}$	$(1 - \alpha L_{N1X})^5\,(1 - \alpha L_{XN2})(1 - \rho_{N_1})^2 (1 - \rho_X)^2\,(1 - \delta_X)$
$N_1 - X - Y - X - N_1 - X - N_2$	$i = 4$	$3L_{N1X} + 2L_{XY} + L_{XN2}$	$(1 - \alpha L_{N1X})^3\,(1 - \alpha L_{XY})^2 (1 - \alpha L_{XN2})(1 - \rho_{N_1})(1 - \rho_Y)\,(1 - \delta_X)^3$
$N_1 - X - N_1 - X - Y - X - N_2$	$i = 5$	$3L_{N1X} + 2L_{XY} + L_{XN2}$	$(1 - \alpha L_{N1X})^3\,(1 - \alpha L_{XY})^2 (1 - \alpha L_{XN2})(1 - \rho_{N_1})(1 - \rho_X)\,(1 - \rho_Y)\,(1 - \delta_X)^2$
$N_1 - X - N_2 - X - N_2$	$i = 6$	$L_{N1X} + 3L_{XN2}$	$(1 - \alpha L_{N1X})\,(1 - \alpha L_{XN2})^3 (1 - \rho_{N_2})\,(1 - \rho_X)\,(1 - \delta_X)$
$N_1 - X - Y - X - Y - X - N_2$	$i = 7$	$L_{N1X} + 4L_{XY} + L_{XN2}$	$(1 - \alpha L_{N1X})\,(1 - \alpha L_{XY})^4 (1 - \alpha L_{XN2})\,(1 - \rho_X)\,(1 - \rho_Y)^2\,(1 - \delta_X)^2$
$N_1 - X - N_1 - X - N_2 - X - N_2$	$i = 8$	$3L_{N1X} + 3L_{XN2}$	$(1 - \alpha L_{N1X})^3\,(1 - \alpha L_{XN2})^3 (1 - \rho_{N_1})\,(1 - \rho_{N_2})\,(1 - \rho_X)^2\,(1 - \delta_X)$
$N_1 - X - N_2 - X - N_1 - X - N_2$	$i = 9$	$3L_{N1X} + 3L_{XN2}$	$(1 - \alpha L_{N1X})^3\,(1 - \alpha L_{XN2})^3 (1 - \rho_{N_1})\,(1 - \rho_{N_2})\,(1 - \delta_X)^3$

Table 9.3 Table of notations

Notation	Definition
ID_{Len}	PL ID length
T_s	Sampling time
Q_{th}	Quantization threshold
N_{Obs}	Number of observations
N_i	Node index i
L_{XY}	The transmission distance of $X - Y$ link
$h_{X,Y}$	The $X - Y$ link CIR
$\hat{h}_{X,Y}$	The estimated $X - Y$ link CIR
L_{min}	The shortest distance between adjacent nodes
d_{max}	The largest distance between adjacent nodes
d_{min}	The shortest distance between a node and the location of a topology change
α	The propagation attenuation factor
ρ_X	The reflection attenuation factor at node X
δ_X	The discontinuity attenuation factor at node X
l_i	ith path length
g_i	ith path gain
$ID^{(0)}_{N_X,N_Y}$	Original PL ID between nodes N_X and N_Y
$ID^{(1)}_{N_X,N_Y}$	Updated PL ID between nodes N_X and N_Y
P_{FND}	The probability of false-negative detection
P_{FPD}	The probability of false-positive detection
P_{SDP}	The probability of successful detection
P_{TC}	The probability of a topology change

- **Step 1**: It involves the channel probing that estimates the different N_{Obs} CIRs from the received signals: $y_{N_i,N^i_j}(n), n \in \{1, ..N_{Obs}\}$.
- **Step 2**: In this step, the channel estimation error is minimized by averaging the estimated CIRs, according to (9.3).
- **Step 3**: Finally, the CIR quantization is performed to provide the PL ID ID_{N_i,N^i_j} of the link N_i-N^i_j.

The generated PL ID here is the primary factor used to recognize alterations in the topology being examined. For example, assume that there is a new connected branch $(X - Y)$, as shown in Fig. 9.1. By using the introduced technique, this connection can be easily detected as follows:

First, node N_2 generates a PL ID $ID^{(0)}_{N_1,N_2}$ for the N_1-N_2 link using Algorithm 3. After a given monitoring period, N_2 should regenerate a new PL ID; $ID^{(1)}_{N_1,N_2}$. Then, if the Bit Mismatch Rate (BMR) between $ID^{(1)}_{N_1,N_2}$ and $ID^{(0)}_{N_1,N_2}$ is greater than a

Algorithm 3 N_i-N_j^i link PL identification scheme [89]

Inputs: PL ID Length: ID_{Len}
Sampling Time: T_S
Quantization Threshold: Q_{th}
Number of Observations: N_{Obs}
Received Signals: $y_{N_i,N_j^i}(n), n \in \{1, ..N_{Obs}\}$

Step 1: Channel Probing:
Estimation of the different N_{Obs} CIRs: $\hat{h}_{N_i,N_j^i}(n)$

Step 2: Minimizing the Channel Estimation Error:

$$\tilde{h}_{N_i,N_j^i} = \frac{\sum_{n=1}^{N_{Obs}} \hat{h}_{N_i,N_j^i}(n)}{N_{Obs}} \tag{9.3}$$

Step 3: CIR quantization to generate the $N_i - N_j^i$ link:
PL ID: ID_{N_i,N_j^i}

predefined threshold, a detected topology change is identified. Otherwise, no such announcement is made.

9.4.2 Topology Change Detection Algorithm

Taking a more general topology into account, as shown in Fig. 9.2, Algorithm 4 introduces the following seven main steps of the multi-node topology change detection system:

- **Step 1:** This step involves the initial signaling between each node N_i and all the other nodes in the considered topology. In particular, N_i should assess the Received Signal Strength Indicators (RSSI)s of all the signals received from the different nodes of the considered topology, where it is assumed that unregistered or unidentified nodes are unable to communicate with registered nodes in this topology.
- **Step 2:** In this step, each node N_i should create a list of N_j^i nodes with a significant RSSI, where $j \in \{1, .., L_i\}$ and N_j^i denote the jth node in N_i's list of length L_i. For nodes that cannot be connected to N_i or have a very low RSSI, the corresponding PL ID cannot be generated or is not precise enough to be taken into account in the proposed algorithm. A high channel estimation error is caused by a very low RSSI, which leads to an inaccurate PL ID.

Algorithm 4 Topology change detection approach [89]

Inputs: PL ID Length: ID_{Len}
Sampling Time: T_S
Quantization Threshold: Q_{th}
Number of Observations: N_{Obs}
Detection Period
Step 1: Initial signaling between each node N_i and the different
other nodes.
Step 2: Each node N_i should generate a list of
N^i_j, $j \in \{1, .., L_i\}$, nodes with significant RSSIs.
Step 3: Each node N_i generates a list of PL IDs:
$ID^{(0)}_{N_i,N^i_j}$, $j \in \{1, .., L_i\}$, using Algorithm 3.
Step 4: Waiting for a detection period, then redo **Steps** 1–3.
to get a new list of PL IDs: $ID^{(1)}_{N_i,N^i_j}$.
Step 5: For each node N_i, select the node N^i_{j*}, with the highest
PL ID variation:
• **for** j from 1 to L_i **do**

$$V_j = \frac{\sum_{m=1}^{ID_{Len}} \left| ID^{(1)}_{N_i,N^i_j}(m) - ID^{(0)}_{N_i,N^i_j}(m) \right|}{\sum_{m=1}^{ID_{Len}} ID^{(0)}_{N_i,N^i_j}(m)}. \tag{9.4}$$

end for
• Select the node N^i_{j*} with the highest PL ID variation.

Step 6: Each node N_i, with $N^i_{j*} \neq \emptyset$, sends its token to N^i_{j*}.

Step 7: If there is a token exchange between two given nodes,
a topology change will be detected on the corresponding
link.

- **Step 3:** Each node N_i should create a list of PL IDs $ID^{(0)}_{N_i,N^i_j}$, $j \in \{1, .., L_i\}$, using Algorithm 3 and the outputs of **Steps 1 and 2**.
- **Step 4:** After a predefined detection period, each node N_i should regenerate a new list of PL IDs: $ID^{(1)}_{N_i,N^i_j}$, while considering the same nodes identified in **Step 2**.
- **Step 5:** In this step, each node N_i should select the node N^i_{j*}, with the highest PL ID variation as presented in Algorithm 4. In particular, each node N_i should evaluate the different PL ID variations (V_j) of the links $N_i - N^i_j$, $j \in \{1, .., L_i\}$. The variation V_j is defined as the ratio of the number of bits that are mismatched

between $ID^{(0)}_{N_i,N^i_j}$ and $ID^{(1)}_{N_i,N^i_j}$ to the total number of $ID^{(0)}_{N_i,N^i_j}$ bits that are 1s. Next, N_i should choose the node with the highest fluctuation V_j.

- **Step 6:** Each node N_i, with $N^i_{j*} \neq \emptyset$, sends its token to node N^i_{j*}.
- **Step 7:** Finally, if two given nodes exchange a token, a topology alteration will be detected on the relevant link, and detection messages will be sent. Otherwise, no topology changes will be noticed.

9.5 Performance Analysis and Numerical Results

In this section, a detailed description of the performance analysis metrics is presented, followed by numerical results and a discussion to investigate the performances of the proposed approach when applied to the two considered system models.

9.5.1 Performance Analysis Metrics

To evaluate the introduced approach's advantages, the following performance analysis metrics have been used:

- The False Negative Detection Probability (FNDP) : denoted by P_{FND}, which is defined as the probability that the proposed technique is unable to detect a topology change.
- The False Positive Detection Probability (FPDP) : denoted by P_{FPD}, which is defined as the probability that the proposed technique detects a nonexistent topology change.
- The Successful Detection Probability (SDP) : denoted by P_{SCD}, which is defined as the probability that the proposed technique correctly detects a topology change and, more specifically, it detects a topology change and identifies the link where the relevant change is located. P_{SDP} can be expressed as follows:

$$P_{SCD} = P_{TC}(1 - P_{FND}) + (1 - P_{TC})(1 - P_{FPD}), \tag{9.5}$$

where P_{TC} denotes the probability of a topology change.

9.5.2 Performance Analysis in a Two-Node Topology Model

Without loss of generality, the conducted simulation parameters are set to be as follows [89]: the distance between N_1 and N_2 is a uniform random variable possible

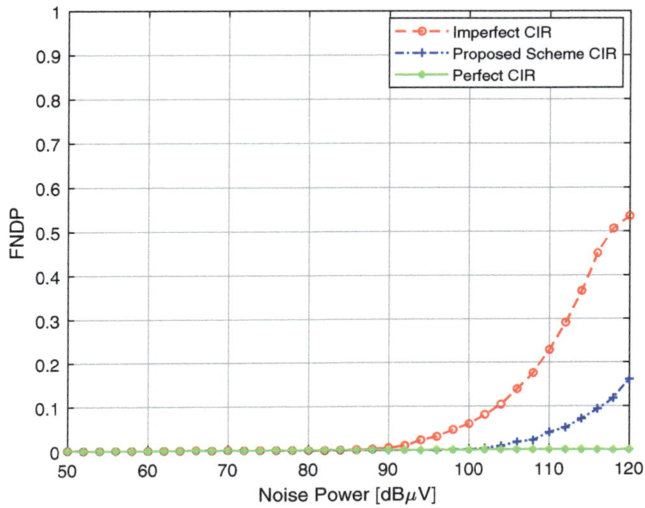

Fig. 9.3 FNDP versus noise power for a two-node topology, with $P_{TC} = 0.5$ [89]

values of $d_{N_1N_2} \in [L_{min}, L_{max}]$, $d_{N_1X} \in [d_{min}, d_{N_1N_2} - d_{min}]$, $d_{XY} \in [L_{min}, L_{max}]$, $L_{min} = 50$ m, $L_{max} = 200$ m, $d_{min} = 10$ m, $\alpha = -30$ dB/m, $\rho_X = -20$ dB, $\delta_X = -10$ dB, the sampling time = 0.005 µs, the PL ID length is equal to 512 bits, $N_{Obs} = 20$, the transmit power is equal to 126 dBµV[110, 111], the range of average noise power is from 50 dBµV to 120 dBµV [130, 204, 226], and the number of iterations is equal to 10^4.

Figure 9.3 presents the FNDP variation versus noise power for the two-node topology, with $P_{TC} = 0.5$. As shown in this figure, the FNDP is directly proportional to the noise power level for the introduced technique with imperfect knowledge of the CIR, which is expected. This is due to the fact that increasing the noise power increases the channel estimation error; hence, the FNDP increases. Furthermore, it is shown that the proposed CIR estimation method provides better performance than that of the considered related work due to the error-minimizing method used within the proposed scheme.

The variations of the FPDP versus the noise power for the two-node topology are illustrated in Fig. 9.4. As depicted in this figure, for low noise power levels, the FPDP is null, which is expected. Then, FPDP increases until it reaches its peak when the noise power levels are increased. This can be explained by the fact that increasing the noise power increases the CIR estimation error, and hence, an increase in the FPDP can be observed. Moreover, when the noise levels increase significantly (e.g., above 90 and above 105), both errors of the CIR estimation and the relevant PL ID increase, which results in a high probability of resemblance between the reference PL ID, $ID^{(0)}_{N_1,N_2}$, and the updated one, $ID^{(1)}_{N_1,N_2}$, leading to a reduction in the FPDP.

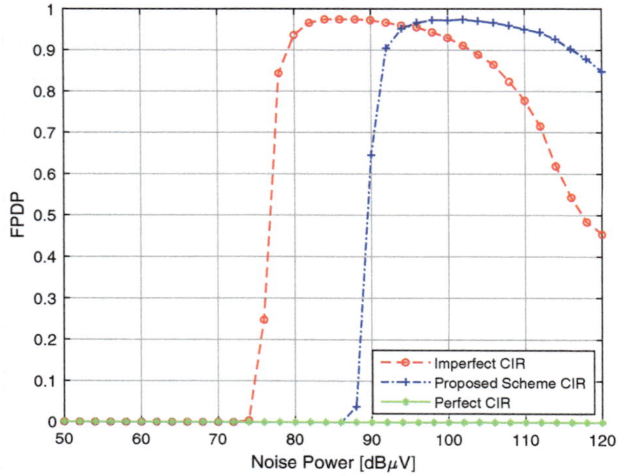

Fig. 9.4 FPDP versus noise power for a two-node topology, with $P_{TC} = 0.5$ [89]

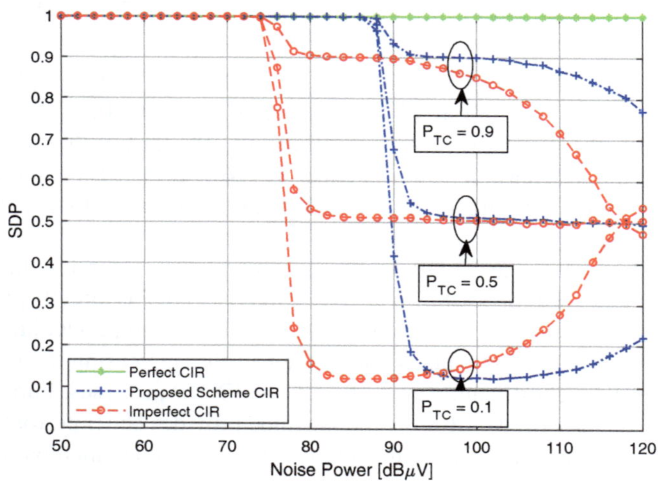

Fig. 9.5 SDP versus noise power for a two-node topology, with $P_{TC} = 0.1,\ 0.5$ and 0.9 [89]

Figure 9.5 presents the SDP variations versus noise power for different values of P_{TC}. For a noise power ($\leq$72 dBμV), the SDP is always equal to one. In fact, it is clear from Figs. 9.3 and 9.4 that both the FNDP and FPDP are zero when the noise is low. Consequently, the SDP is equal to one, according to Eq. (9.5). Then, by increasing the noise power, the FNDP increases, and the FPDP proportionally decreases, resulting in the SDP converging to P_{TC} according to Eq. (9.5). Moreover, when the noise levels increase significantly, the SDP drops to 0.5. This is because the influence of the noise on the PL ID accuracy is rather large in this case, resulting in identical values of the FPDP and FNDP (i.e., 0.5); therefore, the SDP converges to 0.5, according to (9.5). By further increasing the noise power (>120 dBμV), the SDP converges to $1 - P_{TC}$. In particular, for a very high noise level, the FPDP converges to 0 and the FNDP to 1. Hence, based on (9.5), the SDP converges to $1 - P_{TC}$. However, this latter case is uncommon in practice, as noise levels are often less than or equal to 100 dBμV [204, 226].

9.5.3 Performance Analysis in a Multi-Node Topology Model

In this subsection, a multi-node topology model with 14 nodes and 25 discontinuity points has been considered. Without loss of generality, the simulation parameters were set as follows: the distance between every two adjacent points (nodes, discontinuities, or terminals) is a uniform random variable with possible values from $L_{min} = 10$ m to $L_{max} = 50$ m, $\alpha = -30$ dB/m, $\rho_X = -20$ dB, $\delta_X = -10$ dB, $T_s = 0.005$ μs, the transmit power is equal to 126 dBμV, the average noise power range is from 50 to 120 dBμV, the PL ID length is equal to 512 bits, $N_{Obs} = 20$, and the number of iterations is equal to $= 10^3$. Additionally, two unknown connections at X_1, X_2 are added to the considered multi-node topology. To ensure the detection of these connections, P_{TC} was set as one.

The SDP variations versus noise level for the considered scenario of a multi-node topology are presented in Fig. 9.6.

The graph depicts that the SDP is equal to one when the noise power is less or equal to 89 dBμV for the related technique and is less or equal to 100 dBμV for the introduced technique. By increasing the noise level, the SDP converges to zero as the FNDP increases significantly, which is expected. These observations confirm the advantage of the proposed scheme in identifying topology changes in practical PLC environments, with common noise power, e.g., less than/or equal to 100 dBμV [130, 204].

Furthermore, this figure presents the corresponding SDP's Standard Error of the Mean (SEM) for the different considered schemes. This metric measures the degree to which the sample mean differs from the true population mean. Accordingly and as shown in Fig. 9.6, the SEM values are very low compared to the SDP ones, which confirms the advantage of the introduced technique.

9.5.4 Applicability

Reiterating the points made in this book, there are several key elements and underlying conditions that must be taken into account when assessing the suitability of the proposed solution, as well as most signal-based PLC techniques in general. These include the following:

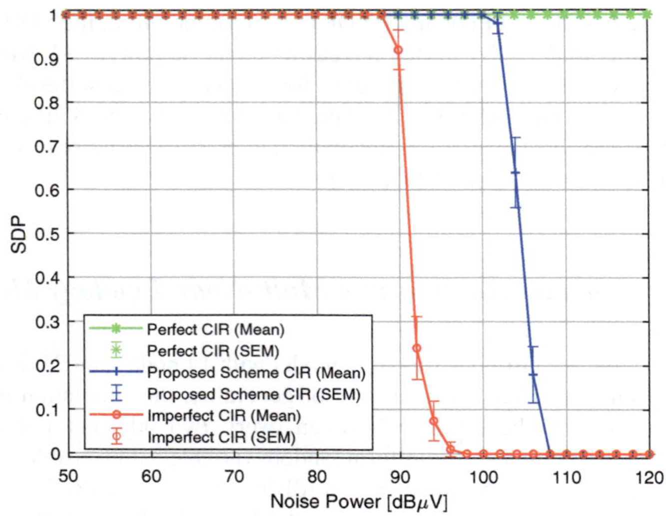

Fig. 9.6 Mean and SEM of SDP versus noise power for a multi-node topology, with two unknown connections [89]

- Alterations in power grids, apart from those in the in-home environment, are usually rare.
- The detection resolution for topology changes is dependent on the employed sampling frequency. Therefore, to achieve an accurate detection, a high sampling frequency is necessary, making the introduced solution more suitable, as a general rule, for Broadband Power Line Communication (BB-PLC) implementations.
- A crucial condition for the applicability of the introduced solution is the maximum length (D_{max}) of the link between two adjacent detectable nodes. In particular, it should be noted that the generation of the PL ID of a connection between two adjacent nodes is dependent upon the particular system and environmental factors, including the length of the PL ID, the sampling time, the link length, and the signal attenuation factors. In addition, the maximum sensing period during the channel probing step between two given nodes is equal to $ID_{Len} \times T_S$. As a consequence, to detect at least the first path of a link between two given nodes, the maximum relevant length must be equal to or less than $D_{max} = ID_{Len} \times T_S \times \varepsilon \times C$, where ε represents the signal propagation speed, e.g., 0.6 [130], and C presents the light speed. Otherwise, no path can be detected, and hence, no PL ID can be generated and used for the proposed detection approach. For instance, for $T_S = 5$ ns and a PL ID length of 512 bits, D_{max} should be equal to 460.8 m, which presents the range that satisfies all the possible applications discussed previously in this chapter.

Chapter 10
PLC Availability Scheme

Abstract Ensuring availability in Power Line Communication (PLC) networks is paramount to maintaining uninterrupted communication, especially given their critical role in critical infrastructure. While most of the existing research has been dedicated to ensuring the availability of these devices, there remains an overlooked area when it comes to Denial of Service (DoS) attacks. Current mechanisms might identify a jammer close to the target devices, yet such devices may encounter difficulties signaling interference under jamming conditions. On the other hand, when these systems are considerably remote from the jammer, detecting its presence becomes quite complicated. This chapter introduces a novel method for detecting the presence of a jammer in a PLC network, which serves to safeguard the availability of remote network equipment. The results of the proposed technique are able to locate a jammer with a probability exceeding 0.99, even when it is positioned 75 m away.

10.1 Denial of Service Attacks in PLC Networks and Jamming Detection Challenges

Given Power Line Communication (PLC)'s instrumental role in critical infrastructures, ensuring its security is imperative. While a substantial body of research in the domain of PLC focuses on issues related to availability—such as latency, throughput, and reliability—there exists a notable gap in the literature concerning denial-of-service attacks. Given the pivotal role that PLC technology plays in contemporary and future sustainable systems, addressing this gap is crucial.

In this setting, Denial of Service (DoS) attacks pose a significant obstacle and challenge to PLC security. Various methods can be employed to disrupt communications and data transmissions among legitimate nodes within the network. One such widely recognized form of DoS attack is jamming . This is a physical layer attack that involves the injection of noise and interference signals into communication channels, thereby inhibiting the nodes from accurately receiving

J. Hernandez Fernandez et al., *Physical Layer Security in Power Line Communications*, Advances in Information Security 108,
https://doi.org/10.1007/978-3-031-57349-1_10

transmitted data. Given these circumstances, the detection of jamming attacks is of paramount importance, although accomplishing this is notably difficult.

Indeed, a malevolent device has the capability to introduce artificial noise into the system, effectively degrading the received Signal-to-Noise Ratio (SNR). This compromises the reliability of the corresponding PLC link in question. In such instances, the entire communication infrastructure may become nonoperational, rendering it incapable of reporting any incidents of jamming detection.

Moreover, traditional jamming detection methods that rely on established metrics such as SNR, Bit-Error Rate (BER), and Packet Error Rate (PER) prove to be ineffective in these circumstances. The challenge lies in the difficulty of differentiating between a genuine communication line failure and a deliberate DoS attack, making these conventional techniques ill-suited for effective jamming detection in PLC systems.

In the literature, many works have been presented in the context of jamming detection for smart grids. Kurt et al. [128] have investigated two different adversary smart grid system models, and they have evaluated the injection of Additive White Gaussian Noise (AWGN) into the communication links to compromise a subset of meters. They have used a Cumulative Sum Control Chart (CUSUM)-based technique to detect the considered attacks. Shin et al. [217] have proposed a solution for reactive jamming attacks for low-cost resource-constrained devices in a smart grid, which consists of isolating the compromised nodes. Another contribution has been presented in [8], where the authors used neural networks to detect and identify jamming attacks. This approach presents a promising solution in this context, especially the use of Deep Learning (DL). However, to the best of our knowledge, jamming detection in PLC networks through DL techniques has not been introduced yet.

This chapter focuses on detecting jamming signals in PLC systems at the physical layer. The introduced contribution consists of an innovative technique that converts physical layer information (I-Q samples) into images. These images will then be processed by Convolutional Neural Network (CNN) to detect potential jamming attacks.

10.2 Adversary Model and Motivation

This section presents the considered adversary model in this chapter, along with the relevant motivation. The system model consists of a house, where two legitimate PLC devices, $\mathcal{A}$ and $\mathcal{B}$, communicate via the power line network in the presence of a malicious user ($\mathcal{E}$), as illustrated in Fig. 10.1. In this chapter, we focus on accessibility challenges, where $\mathcal{E}$ would like to prevent communication between $\mathcal{A}$ and $\mathcal{B}$, using strategies to carry out a DoS attack.

A DoS attack can have a major effect on PLC due to the critical nature of many services that rely on these systems. In the following, some potential targets for an adversary attack are presented:

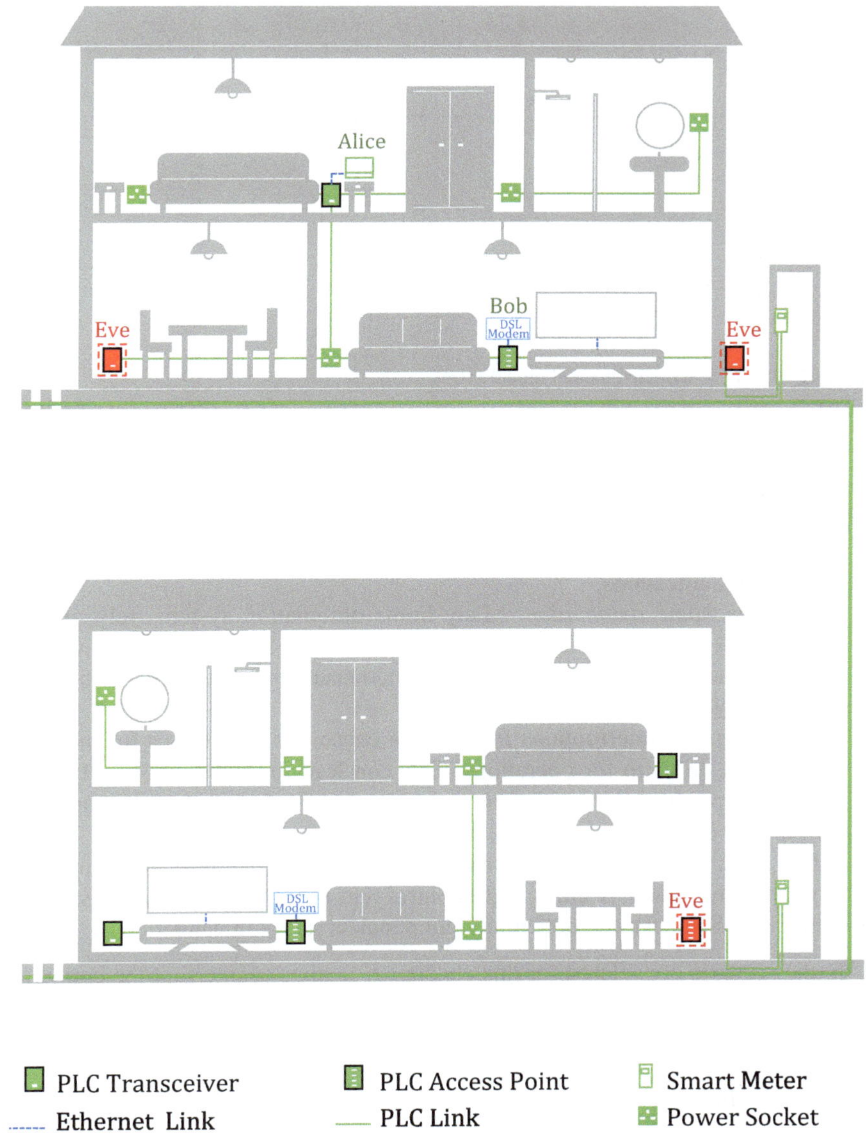

Fig. 10.1 The adversary system model [109]

- *Interruptions in energy services.* The management of energy distribution is a key element of smart grid systems. Therefore, if the related communications are hindered, it could cause a disruption of these services, resulting in a power outage.

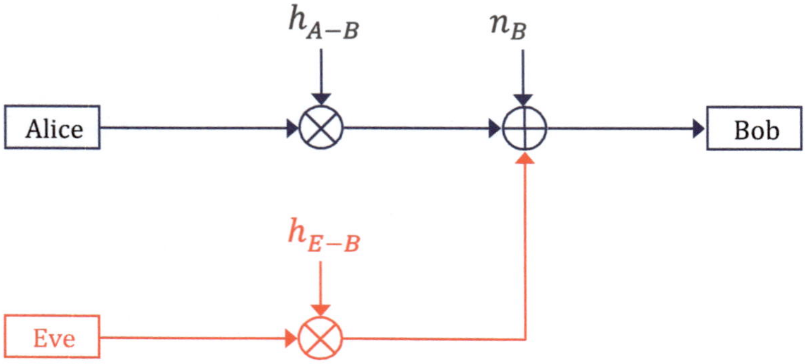

Fig. 10.2 The PLC wiretap channel model [109]

- *Communications disruption.* Although wireless communications and PLC have the same challenges in avoiding the DoS attacks, damages in the PLC network could potentially interrupt communications over a wider area, affecting different industries and services, such as Internet Service Providers (ISP)s and public safety communications.
- *Impacts on home automation.* Many home automation systems rely on PLC technologies for communication between devices. DoS attacks can disrupt these systems, leading to inconvenience for homeowners and potentially compromising safety systems.
- *Financial losses.* Interruptions in PLC can cause financial losses for industries that depend on them for operations. This can lead to downtime, resulting in a decrease in productivity and income, as well as additional expenses to strengthen the systems against future attacks.
- *Safety risks.* PLC can be deployed to carry out safety-related tasks in crucial infrastructures, such as traffic light control and hospital power systems. In this context, DoS attacks can potentially put people's safety in danger.

In the scenario considered for this chapter, Eve aims to interrupt the communications between $Alice$ and Bob, by injecting an interference signal into the PLC network. Accordingly, the corresponding wiretap channel model is illustrated in Fig. 10.2, where h_{A-B} and h_{A-E} are the complex channel gains of the link $Alice$-Bob and the link $Alice$-Eve, respectively, while n_B is the relevant additive noise.

In view of this, the instantaneous received SNR at B can be expressed as follows:

$$\gamma_{A-B} = \frac{P_T \, H_{A-B}}{P_J \, H_{E-B} + N_B}, \tag{10.1}$$

where P_T is the transmit signal power, P_J is the jamming signal power, $H_{X-B} = |h_{X-B}|^2$, $X \in \{A, E\}$, and N_B is the relevant noise power.

Table 10.1 Main notation and values

Parameter	Notation	Value
Transmit power [dBμV]	P_T	120
Noise power [dBμV]	P_N	70
Nbr. of SC per OFDM symbol	N_{SC}	512
FFT Nbr.	N_{FFT}	512
Carrier frequency [Hz]	F_c	$20\ 10^6$
Subcarrier spacing [Hz]	Δ_f	$15\ 10^3$
Nbr. of OFDM symbols per frame	N_{Symb}^{Frm}	20
Sampling time [s]	T_{samp}	$1/(2 \times F_c)$
OFDM symbol duration [s]	T_{Symb}	$N_{FFT} \times T_{samp}$
Frame duration [s]	T_{Frame}	$N_{Symb}^{Frm} \times T_{Symb}$
Total Nbr. of frames	N_{Frame}	$\lceil T_{Sim} / T_{Frame} \rceil$
Total Nbr. of OFDM symbols	N_{Symb}	$N_{Frame} \times N_{Symb}^{Frm}$
Distance *Alice- Bob* [m]	d_{A-B}	30
Distance *Eve–Bob* [m]	d_{E-B}	30
Simulation time [s]	T_{Sim}	2

In this work, H_{X-B} is assumed to be an exponential random variable, with Probability Distribution Function (PDF) given by Mathur et al. [151], Abou-Rjeily [1], Ai and Cheffena [7]

$$f_{H_{X-B}}(x) = \frac{1}{\bar{H}_{X-B}(F_C, d_{X-B})} \exp\left(\frac{-x}{\bar{H}_{X-B}(F_C, d_{X-B})}\right), \tag{10.2}$$

where $\bar{H}_{X-B}(F_C, d_{X-B})$ is the mean of H_{X-B}, F_C is the used carrier frequency, and d_{X-B} is the corresponding distance of the $X-B$ link. Based on [7], $\bar{H}_{X-B}(F_C, d_{X-B})$ is expressed as

$$\bar{H}_{X-B}(F_C, d_{X-B}) = \exp\Big(- 2 \left[a_0 + a_1\ F_C^{a_2}\right] d_{X-B}\Big), \tag{10.3}$$

where a_0, a_1, and a_2 are the PLC environment-based channel attenuations.

Accordingly, the main notations used in this chapter, along with their associated values, are presented in Table 10.1.

10.3 Methodology

This section details the introduced jamming detection technique. The main idea consists of transforming the I-Q samples of the received signals into images.

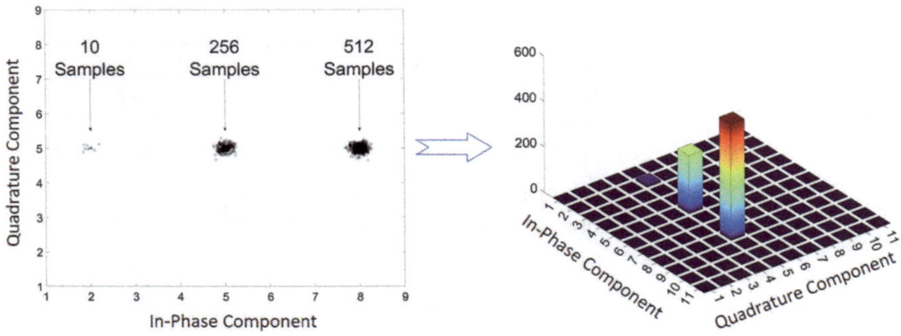

Fig. 10.3 Illustration of an example of the bivariate histogram results [109]

Then, an image classification approach should be used to detect which images (and correspondingly, I-Q samples) are affected by jamming.

In particular and without loss of generality, the Binary Phase Shift Keying (BPSK) modulation is considered in this chapter. This modulation involves a null imaginary component ($I = 0$) and a real component Q equals to -1 or 1, depending on the input binary symbol b, i.e.,

$$b = 0 \rightarrow Q = 1, I = 0.$$
$$b = 1 \rightarrow Q = -1, I = 0. \tag{10.4}$$

The first step of the proposed technique consists of converting these I-Q samples into images. In this context, many research attempts and relevant works have been presented in the literature to generate Red Green Blue (RGB) images from I-Q samples [8, 9, 173]. In this work, the state-of-the-art solutions are readopted to maximize the relevant performance of the considered use case.

Based on empirical considerations, 1,597,440 I-Q samples per image and a total number of 1500 images for each experiment (750 jamming and 750 no jamming) have been considered in this work.

To further explain the introduced jamming detection approach, Fig. 10.3 summarizes the overall approach process. First, the I-Q plane is split into tiles, i.e., 224×224. Then, a bivariate histogram is computed. For each tile, the number of I-Q samples should be computed and mapped into a pixel value. For instance, Fig. 10.3 depicts 3 clouds of I-Q samples with 10, 256, and 512 samples, placed on 3 different tiles at positions [2, 5], [5, 5], and [8, 5]. It is worth noting that the cloud positions are not meaningful, and they are chosen only for the sake of clarity. The mentioned bivariate histogram counts the number of I-Q samples per tile, i.e., 10, 256, and 512, respectively. Then, the output should be considered as the value of a pixel in the image. Hence, by assuming that h is the output of the bivariate histogram, the pixel coordinates p_R, p_G, and p_B should be given by

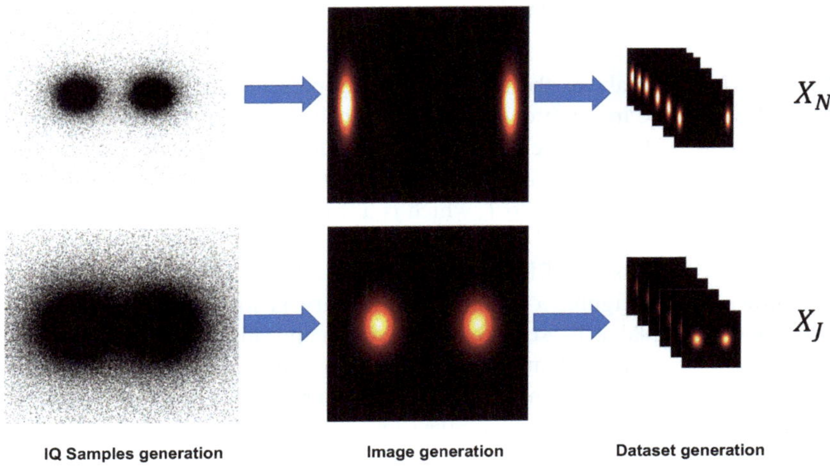

Fig. 10.4 I-Q samples are converted to images, which in turn are organized into two datasets, i.e., X_{NJ} and X_J [109]

- $0 \leq h \leq 255$, then $p_R = 0$, $p_G = 0$, $p_B = h$,
- $256 \leq h \leq 511$, then $p_R = 0$, $p_G = h - 255$, $p_B = 255$,
- $h > 511$, then $p_R = h - 510$, $p_G = 255$, $p_B = 255$.

In addition, it was observed that when $h > 767$, the output is clipped—this problem can be solved by accurately setting the amount of samples in the I-Q plane, as previously mentioned.

As shown in Fig. 10.4, the image generation process is summarized. First, the I-Q samples should be generated according to the model proposed in Sect. 10.2. Then, the output should be converted into images, and finally, two *datastores* of images will be generated, i.e., X_J and X_{NJ}, including images from the jammed and not-jammed communication links, respectively.

The generated *datastores*, X_J, and X_{NJ}, are the input for the introduced deep learning approach that is based on the classification process. More specifically, each datastore X_y, $y \in \{J, NJ\}$, should be split into 3 subsets of sizes 0.6, 0.2, and 0.2, compared to the original one, i.e., $\mathcal{T}, \mathcal{V}, \mathcal{S}$, which are considered for training, validation, and testing, respectively.

Then, by considering a state-of-the-art DL classifier (CNN) based on a Residual Network (Resnet-18) already implemented in MatLab2022b, the neural network model is pre-trained with images from the ImageNet database [208], thus requiring some modifications. Accordingly, the input layers were adjusted to fit the output of the bivariate histogram (image size) of 224×224, and the output layer was modified to fit the two classes of our problem, *jamming* and *non-jamming*. The model was then partially retrained to incorporate the features of the images created from the I-Q samples.

10.4 Performance Analysis

In this performance analysis, the adversary model presented in Fig. 10.2 has been considered, where the transmitter ($\mathcal{A}$) and the receiver ($\mathcal{B}$) are set to be 30 meters apart. Additionally, the distance between $\mathcal{E}$ (Jammer) and $\mathcal{B}$ is set to range from 5 to 60 meters, as shown in Fig. 10.5. Also, different relative jamming transmission powers are considered from 0.1 to 1, which is a fraction of the transmission power of $\mathcal{A}$ (120 dBμV).

Figure 10.5 presents the BER variation versus the distance between $\mathcal{E}$ and $\mathcal{B}$. As shown in this figure, a theoretical boundary is identified, e.g., the used jamming power is equal to 1—solid red line—which presents the scenario where the jammer is using the same transmit power of the legitimate transmitter. This scenario highlights the case of an adversary using the same hardware as the legitimate one—in this situation, an adversary could use the same commercial devices or gain control of one of the already deployed devices. It is noteworthy that when the jamming power is equal to the legitimate transmitter's power and the jammer is at the same distance, the BER is 0.15. It is also shown that the jamming power has a significant impact on the BER, meaning that decreasing the jamming power will reduce the BER regardless of the distance between the jammer and the receiver.

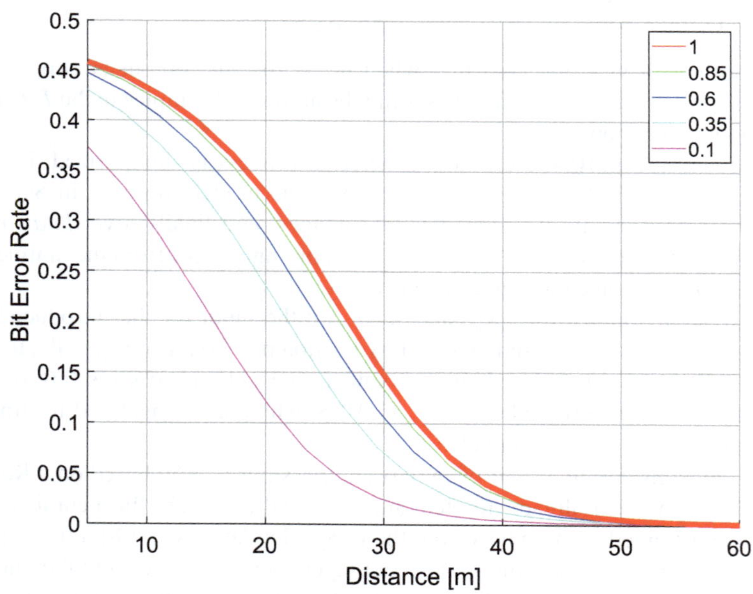

Fig. 10.5 The BER variations versus the distance between the legitimate receiver and the jammer, with $P_J = a\ P_T,\ a \in \{1, 0.85, 0.6, 0.35, \text{and } 0.1\}$ [109]

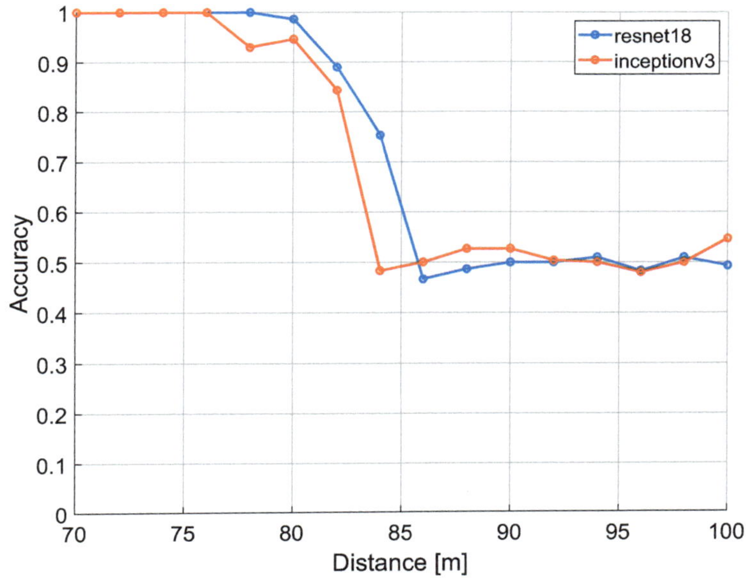

Fig. 10.6 The variations of the proposed techniques' accuracy versus the distance between the legitimate receiver and the jammer, with $P_J = P_T$ [109]

Finally, it is worth mentioning that, in this considered scenario, the jammer impacts the legitimate receiver up to a distance of 50 m when the used jamming power is less or equal to the legitimate transmission power.

This is a limited point to traditional detection methods that rely on upper-layer jamming detection metrics such as signal-to-noise ratio and BER. In fact, when the jammer is connected at a given location, all devices in the neighborhood (less than 50 m) can potentially detect its presence, but they cannot communicate it, as the power line medium is jammed. On the contrary, devices located more than 50 m away from the jammer cannot detect it, as the BER is nearly zero, and the related metrics do not indicate its presence.

On the other hand, the introduced contribution has been designed to detect the presence of a jammer far away from legitimate devices, i.e., when the distance between the receiver and the jammer is greater than 50 m, in this scenario. Accordingly, Fig. 10.6 presents the introduced classifier's results, with different values of the distance between the legitimate receiver and the jammer. The results confirm the capability of the proposed technique to detect the jamming attack with a probability of >0.99 within a jamming distance of 75 m. It is worth mentioning that the neural networks (*resnet18* and *inceptionv3*) offer the same results. However, *resnet18* is slightly more robust to channel noise when the distance is greater than 75 m.

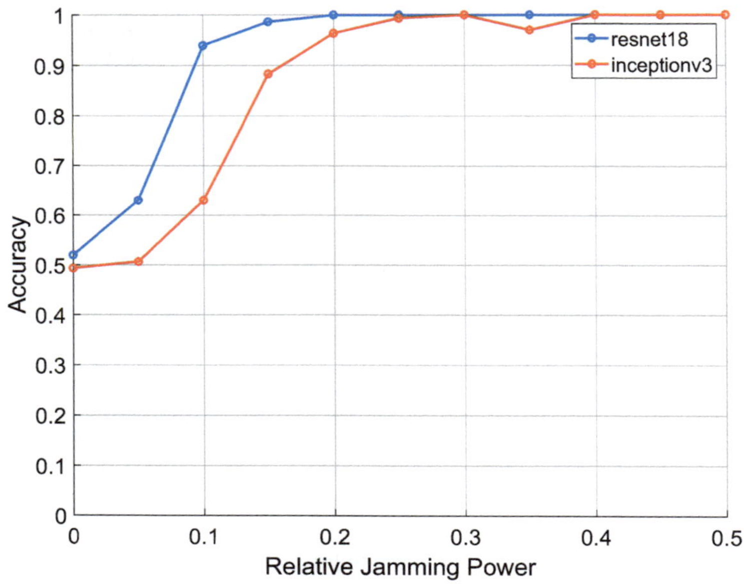

Fig. 10.7 The variations of the proposed techniques' accuracy versus the used jamming power, with the distance between the legitimate receiver and the jammer is equal to 70 m [109]

The accuracy of the introduced technique is depicted in Fig. 10.7 when the distance between the jammer and the receiver is set to 70 m, and the relative jamming power ranges from 0 to 0.5, which is equivalent to no jamming and half of the legitimate transmit power, respectively. As shown in this figure, the introduced technique is able to detect the presence of a jammer with a probability of more than 0.95 when $P_J > 0.2$. Both neural networks have similar performance, although *resnet18* is more robust to channel noise and jamming.

10.5 Discussion

Jamming a PLC link is just as simple as it is for a wireless connection and has the same damaging effect on both systems—the receiving devices are connected to the power line, and unlike the wireless situation, they may have limited options to protect against the attack. Much attention has been paid to the detection of jamming in radio frequency systems, while only a few studies have focused on PLC systems.

This chapter has highlighted the limitations of current solutions while shedding light on future challenges. In particular, it is challenging to identify jamming in PLC systems since the jammer's primary goal is to obstruct communication. Consequently, while PLC devices may be capable of detecting a jammer's presence, they are incapacitated in communicating this vital information through the same compromised PLC system. This renders taking effective countermeasures virtually impossible. Our proposed solution transcends traditional jamming detection techniques by offering a methodology that can confirm the presence of a jammer in areas where the jammer's impact on the power line is negligible, as indicated by a BER close to zero. With respect to the considered adversary system model, each PLC device can observe the current condition of the medium and detect the presence of a jammer from a far distance (approximately 80 m) by just collecting I-Q samples from the power line to detect the presence of a jammer.

Considering the detrimental effects of jamming on Power Line Communication (PLC) systems, we posit that our proposed approach–utilizing in-phase and quadrature (I-Q) samples converted into image formats in conjunction with Convolutional Neural Networks (CNN)–serves as a promising foundation for future investigations in this domain.

Chapter 11
Conclusions

Abstract In this book, we have explored the critical need for robust security in power line communications (PLC) and PLC's role in our interconnected power grid infrastructure. We also highlight physical layer security (PLS) as a crucial approach for securing low-complexity device networks in PLC applications.

While the initial chapters lay a foundational understanding of the PLC ecosystem, discussing its evolution, technological landscape, and significance, subsequent chapters frame channel characteristics and address various security challenges, introducing the concept of PLS as a transformative approach to safeguarding data transmission. From Chaps. 5–10, the focus shifts to detailed methodologies, performance metrics, and practical implementations that enhance the security at the physical layer.

Despite extensive research and proposed solutions, the book cannot but conclude with a call for further investigation into securing PLC networks. We also suggest future directions for research with actionable impact that could support broader domains such as the Internet of Things (IoT) and edge computing, highlighting the dynamic and evolving nature of the field.

As declared in the Preface, the objective of this book was to tame the complexities, highlight the vulnerabilities, and point to innovative solutions at the intersection of power systems and communication technologies. To this end, the journey undertaken in the preceding chapters has been marked by a thorough exploration of Power Line Communication (PLC) networks, channel characteristics, security threats, and countermeasures, with results supported by rigorous formalism, precise notation, and extensive experimental results. The exposed solutions have also benefited from a contextualization tied to compelling use cases. We can itemize the essence of the topics touched in the previous chapters as follows:

- **Understanding the PLC ecosystem**: This introductory chapter laid a solid foundation by explaining the evolution, pervasiveness, and significance of PLC in modern infrastructure. Acknowledging power lines as a vulnerable environment underscored the criticality of implementing robust security measures.

J. Hernandez Fernandez et al., *Physical Layer Security in Power Line Communications*, Advances in Information Security 108,
https://doi.org/10.1007/978-3-031-57349-1_11

- **Navigating the technological landscape**: The provided overview of PLC networks examines the standardization environment, bandplans, technologies, and protocols, equipping the readers with a holistic understanding of the technological landscape. The diverse applications of PLC, from data transmission to grid information inference, highlighted its versatility and importance.
- **Deciphering channel characteristics**: Chap. 3 explored the key features of PLC channel characteristics, ranging from transmission line theory to multipath propagation and channel noise. The detailed examination of the channel impulse response (CIR) enriched our comprehension of the underlying communication medium and provided a fundamental tool for the solutions presented in the following.
- **Addressing security challenges**: Chap. 4 reviews PLC security, highlighting issues, limitations, and various security threats. The classification of threat sources and types laid the groundwork for the subsequent chapters, which proposed innovative solutions to fortify the security posture of PLC networks.
- **The emergence of physical layer security**: Chaps. 5 through 10 marked a pivotal shift towards the exploration of physical layer security in PLC. These chapters introduced readers to physical layer security concepts, techniques, and methodologies, showcasing the potential to mitigate security threats at the foundational level. Moreover, these chapters provided a comprehensive examination of performance metrics, algorithms, and numerical results pertaining to various aspects of physical layer security. The discussion on confidentiality techniques, authentication methods, network integrity solutions, and availability schemes, other than being formal, precise, and supported by mathematical proof and experiments, also offered practical insights into real-world implementations.

In conclusion, this book serves as a beacon for researchers, engineers, and practitioners working or interested in the evolving landscape of physical layer security in PLC. The blending of theoretical foundations, technological insights, and practical solutions presented within these pages aims at fostering a deeper understanding of the challenges and opportunities inherent in securing the critical intersection of energy grid and communication.

However, despite the efforts of the authors, we recognize that the field is in continuously evolving and that the solution highlighted so far, while sound, viable, and reliable, is only a small step towards the design and implementation of a resilient and secure PLC ecosystem that, once realized, would open up the opportunity to support further domains, like, for instance, IoT and edge cloud. Hence, this book cannot conclude but with a call to pursue further research in the field of physical layer security in PLC.

References

1. Abou-Rjeily, C.: Power line communications under rayleigh fading and nakagami noise: novel insights on the MIMO and multi-hop techniques. IET Commun. **12**(2), 184–191 (2018)
2. ACER/CEER: Annual report on the results of monitoring the internal electricity and natural gas markets in 2021 (2022). https://www.acer.europa.eu/Publications/MMR_2021_Energy_Retail_Consumer_Protection_Volume.pdf
3. Ahiadormey, R.K., Anokye, P., Jo, H.S., Song, C., Lee, K.J.: Secrecy outage analysis in NOMA power line communications. IEEE Commun. Lett. **25**(5), 1448–1452 (2021)
4. Ahmed, M.O., Lampe, L.: Parametric and nonparametric methods for power line network topology inference. In: 2012 IEEE International Symposium on Power Line Communications and Its Applications, pp. 274–279 (2012)
5. Ahmed, M.O., Lampe, L.: Power line network topology inference using frequency domain reflectometry. In: 2012 IEEE International Conference on Communications (ICC), pp. 3419–3423 (2012)
6. Ahmed, M.O., Lampe, L.: Power line communications for low-voltage power grid tomography. IEEE Trans. Commun. **61**(12), 5163–5175 (2013)
7. Ai, Y., Cheffena, M.: Capacity analysis of PLC over rayleigh fading channels with colored nakagami-m additive noise. In: 2016 IEEE 84th Vehicular Technology Conference (VTC-Fall), pp. 1–5 (2016)
8. Alhazbi, S., Sciancalepore, S., Oligeri, G.: Bloodhound: Early detection and identification of jamming at the phy-layer. In: 2023 IEEE 20th Consumer Communications & Networking Conference (CCNC), pp. 1033–1041 (2023)
9. Alhazbi, S., Sciancalepore, S., Oligeri, G.: The day-after-tomorrow: On the performance of radio fingerprinting over time. arXiv preprint arXiv:2305.05285 (2023)
10. Aliberti, G., Di Pietro, R., Guarino, S.: Reliable and perfectly secret communication over the generalized ozarow-wyner's wire-tap channel. Comput. Netw. **109**, 21–30 (2016). https://doi.org/10.1016/j.comnet.2016.06.034
11. Amini, S., Pasqualetti, F., Mohsenian-Rad, H.: Dynamic load altering attacks against power system stability: attack models and protection schemes. IEEE Trans. Smart Grid **9**(4), 2862–2872 (2018)
12. Antoniali, M., Tonello, A.M., Lenardon, M., Qualizza, A.: Measurements and analysis of PLC channels in a cruise ship. In: 2011 IEEE International Symposium on Power Line Communications and Its Applications, pp. 102–107 (2011)
13. Anttiroiko, A.V.: Electronic Government: Concepts, Methodologies, Tools, and Applications: Concepts, Methodologies, Tools, and Applications, vol. 3. IGI Global, Pennsylvania (2008)

J. Hernandez Fernandez et al., *Physical Layer Security in Power Line Communications*, Advances in Information Security 108,
https://doi.org/10.1007/978-3-031-57349-1

14. Aouichak, I., Khalil, K., Elfeki, I., Le Bunetel, J., Raingeaud, Y.: Topology identification method for unknown indoor plc home networks. In: 2017 International Symposium on Electromagnetic Compatibility - EMC EUROPE, pp. 1–4 (2017)
15. Applegate, S.: Cybermilitias and political hackers: use of irregular forces in cyberwarfare. IEEE Security Privacy **9**(5), 16–22 (2011)
16. Arshad, M., Islam, S.M.: A novel fuzzy logic technique for power transformer asset management. In: Conference Record of the 2006 IEEE Industry Applications Conference Forty-First IAS Annual Meeting, vol. 1, pp. 276–286 (2006)
17. Artale, G., Cataliotti, A., Cosentino, V., Guaiana, S., Di Cara, D., Panzavecchia, N., Tinè, G., Fiorelli, R.: Narrowband power line communications for medium voltage smart grids. In: 2014 IEEE International Conference on Smart Grid Communications (SmartGridComm), pp. 469–474. IEEE, Piscataway (2014)
18. Bari, A., Jiang, J., Saad, W., Jaekel, A.: Challenges in the smart grid applications: an overview. Int. J. Distrib. Sensor Netw. **10**(2), 974682 (2014)
19. Ben Hamida, S.T., Pierrot, J., Castelluccia, C.: An adaptive quantization algorithm for secret key generation using radio channel measurements. In: 2009 3rd International Conference on New Technologies, Mobility and Security, pp. 1–5 (2009)
20. Bencsáth, B., Pék, G., Buttyán, L., Felegyhazi, M.: The cousins of stuxnet: Duqu, flame, and gauss. Future Int. **4**(4), 971–1003 (2012)
21. Berger, L. T., Schwager, A., Escudero-Garzás, J.: Power line communications for smart grid applications. J. Electr. Comput. Eng. **2013**, 3 (2013)
22. Berger, L.T., Schwager, A., Pagani, P., Schneider, D.: MIMO Power Line Communications: Narrow and Broadband Standards, EMC, and Advanced Processing. CRC Press, Boca Raton (2014)
23. Berkman, W.H., Mollenkopf, J.D.: Power line communications device and method (2009). US Patent 7,626,489
24. Bisharat, G.E., et al.: Israel's invasion of gaza in international law'(2009–2010) 38. Denver J. Int. Law Policy **38**, 41 (2009)
25. Bobric, G.D.: Study regarding the cyber threats to the national security. Sci. Bull.-Nicolae Balcescu Land Forces Acad. **25**(1), 18–25 (2020)
26. Bou-Harb, E., Fachkha, C., Pourzandi, M., Debbabi, M., Assi, C.: Communication security for smart grid distribution networks. IEEE Commun. Mag. **51**(1), 42–49 (2013)
27. Brusilovsky, A., Faynberg, I., Zeltsan, Z., Patel, S.: Password-authenticated key (pak) diffie-hellman exchange. RFC 5683 (2010)
28. Bucci, G., D'Innocenzo, F., Dolce, S., Fiorucci, E., Ciancetta, F.: Power line communication, overview of standards and applications. In: 21st IMEKO World Congress on Measurement in Research and Industry. IMEKO-International Measurement Federation Secretariat (2015)
29. Bumiller, G., Lampe, L., Hrasnica, H.: Power line communication networks for large-scale control and automation systems. IEEE Commun. Mag. **48**(4), 106–113 (2010)
30. Busquet, A.R.: G. hnem for AMI and DR. In: 2012 International Conference on Computing, Networking and Communications (ICNC). IEEE, Piscataway (2012)
31. Button, M.: Economic and industrial espionage. Security J. **33**, 1–5 (2020)
32. Camponogara, A., Poor, H.V., Ribeiro, M.V.: PLC systems under the presence of a malicious wireless communication device: physical layer security analyses. IEEE Syst. J. **14**(4), 4901–4910 (2020)
33. Camponogara, A., Poor, H.V., Ribeiro, M.V.: Physical layer security of in-home PLC systems: analysis based on a measurement campaign. IEEE Syst. J. **15**(1), 617–628 (2021)
34. Cano, C., Pittolo, A., Malone, D., Lampe, L., Tonello, A.M., Dabak, A.G.: State of the art in power line communications: from the applications to the medium. IEEE J. Sel. Areas Commun. **34**(7), 1935–1952 (2016)
35. Caprolu, M., Hernandez Fernandez, J., Alassi, A., Di Pietro, R.: Increasing renewable generation feed-in capacity leveraging smart meters. In: 2020 IEEE Green Energy and Smart Systems Conference (IGESSC), pp. 1–7 (2020)

36. Caraccilo, D.J., Rohling, A.M.: Targeting in postconflict operations in iraq. Military Rev. **84**(1), 11 (2004)
37. Çelebi, H.B.: Noise and multipath characteristics of power line communication channels [ph. d. Thesis]. University of South Florida, Tampa, FL (2010)
38. Çelebi, H.B., Güzelgöz, S., Güzel, T., Arslan, H.: Noise and channel statistics of indoor power line networks. In: 2011 18th International Conference on Telecommunications, pp. 523–527 (2011)
39. Cerrudo, C.: An emerging US (and world) threat: cities wide open to cyber attacks. Secur. Smart Cities **17**, 137–151 (2015)
40. Chen, P., Liu, X., Zhang, J., Yu, C., Pu, H., Yao, Y.: Improvement of prime protocol based on chaotic cryptography. In: 2019 22nd International Conference on Electrical Machines and Systems (ICEMS), pp. 1–5. IEEE, Piscataway (2019)
41. Cho, M., Huang, H., Chen, C., Thom, H.T., Wang, P., Chang, W., Wang, C.: The implementation and applications of low voltage distribution line theft supervisory system. In: 2016 3rd International Conference on Green Technology and Sustainable Development (GTSD), pp. 178–184 (2016)
42. Christopher, A.V., Swaminathan, G., Subramanian, M., Thangaraj, P.: Distribution line monitoring system for the detection of power theft using power line communication. In: 2014 IEEE Conference on Energy Conversion (CENCON), pp. 55–60 (2014)
43. Chu, G., Li, J., Liu, W.: Narrow band power line channel characteristics for low voltage access network in China. In: 2013 IEEE 17th International Symposium on Power Line Communications and Its Applications, pp. 297–302 (2013)
44. Ciapessoni, E., Cirio, D., Kjølle, G., Massucco, S., Pitto, A., Sforna, M.: Probabilistic risk-based security assessment of power systems considering incumbent threats and uncertainties. IEEE Trans. Smart Grid **7**(6), 2890–2903 (2016)
45. Collins, S., McCombie, S.: Stuxnet: the emergence of a new cyber weapon and its implications. J. Polic. Intell. Counter Terror. **7**(1), 80–91 (2012)
46. CORINEX: PRIME Alliance standardises BPL smart meter gateway solution for management of complex electric grids. https://www.corinex.com/news/bpl-smart-meter-gateway-solution
47. Crane, A.: In the company of spies: when competitive intelligence gathering becomes industrial espionage. Business Horiz. **48**(3), 233–240 (2005)
48. Daş, R., Karabade, A., Tuna, G.: Common network attack types and defense mechanisms. In: 2015 23nd Signal Processing and Communications Applications Conference (SIU), pp. 2658–2661. IEEE, Piscataway (2015)
49. de Leon, D.C., Alves-Foss, J., Krings, A., Oman, P.: Modeling complex control systems to identify remotely accessible devices vulnerable to cyber attack. In: ACM Workshop on Scientific Aspects of Cyber Terrorism, (SACT). Citeseer (2002)
50. Degardin, V., Junqua, I., Lienard, M., Degauque, P., Bertuol, S.: Theoretical approach to the feasibility of power-line communication in aircrafts. IEEE Trans. Vehic. Technol. **62**(3), 1362–1366 (2013)
51. Di Bert, L., Caldera, P., Schwingshackl, D., Tonello, A.M.: On noise modeling for power line communications. In: 2011 IEEE International Symposium on Power Line Communications and Its Applications, pp. 283–288 (2011)
52. Diwold, K., Stifter, M., Zehetbauer, P.: Network and feeder assignment of smart meters based on communication and measurement data. In: 2015 International Symposium on Smart Electric Distribution Systems and Technologies (EDST), pp. 541–546. IEEE, Piscataway (2015)
53. Donovan, G.T.: Russian operational art in the russo-georgian war of 2008. US Army War College (2009)
54. Dudek, S.: HomePlugav PLC: Practical attacks and backdooring (2015)
55. El Mrabet, Z., Kaabouch, N., El Ghazi, H., El Ghazi, H.: Cyber-security in smart grid: survey and challenges. Comput. Electr. Eng. **67**, 469–482 (2018)
56. Elrod, C.T., Kashyap, P.: Network threat detection and mitigation (2012). US Patent 8,255,996

57. Erseghe, T., Lorenzon, F., Tomasin, S., Costabeber, A., Tenti, P.: Distance measurement over PLC for dynamic grid mapping of smart micro grids. In: 2011 IEEE International Conference on Smart Grid Communications (SmartGridComm), pp. 487–492 (2011)
58. Erseghe, T., Tomasin, S., Vigato, A.: Topology estimation for smart micro grids via powerline communications. IEEE Trans. Signal Process. **61**(13), 3368–3377 (2013)
59. Eurelectric: Power distribution in europe - facts & figures (2013). https://cdn.eurelectric.org/media/1835/dso_report-web_final-2013--030-0764-01-e-h-D66B0486.pdf
60. European Parliament: Directive 2014/30/eu of the european parliament and of the council of 26 february 2014 on the harmonisation of the laws of the member states relating to electromagnetic compatibility (2014). http://data.europa.eu/eli/dir/2014/30/oj
61. Fernandez, J.H., Omri, A., Di Pietro, R.: Performance analysis of physical layer security in power line communication networks. In: 2023 IEEE International Symposium on Computers and Communications (ISCC), pp. 777–782 (2023)
62. Ferreira, H.C., Lampe, L., Newbury, J., Swart, T.G.: Power Line Communications: Theory and Applications for Narrowband and Broadband Communications Over Power Lines. Wiley, Hoboken (2011)
63. Fliss, M.R., Hernandez Fernandez, J., Omri, A., Oligeri, G.: NB-PLC successful transmission probability analysis. In: 2nd International Conference on Smart Grid and Renewable Energy (SGRE) (2019)
64. Frenzel, L.E.: Power-line communications emerge as a core networking technology-technology report-as concerns about interference, noise, and attenuation are addressed, new standards expand plc's role in the home and industry. Electron. Des. **60**(7), 38 (2012)
65. G3-PLC Alliance: G3 PLC - enabling the smart grid together. Technical Report, G3-PLC Alliance (2019). http://www.g3-plc.com
66. Galli, S., Scaglione, A., Wang, Z.: For the grid and through the grid: the role of power line communications in the smart grid. Proc. IEEE **99**(6), 998–1027 (2011)
67. Gao, J., Xiao, Y., Liu, J., Liang, W., Chen, C.P.: A survey of communication/networking in smart grids. Future Gener. Comput. Syst. **28**(2), 391–404 (2012)
68. Girotto, M., Tonello, A.M.: EMC regulations and spectral constraints for multicarrier modulation in PLC. IEEE Access **5**, 4954–4966 (2017)
69. Giroux, J.: Targeting energy infrastructure: examining the terrorist threat in North Africa and its broader implications. Elcano Newslett. **0**(53), 10 (2009)
70. Goel, S., Hong, Y.: Security challenges in smart grid implementation. In: Smart Grid Security, pp. 1–39. Springer, Berlin (2015)
71. Goel, S., Bush, S.F., Bakken, D., et al.: IEEE Vision for Smart Grid Communications: 2030 and Beyond. IEEE Standard Association, pp. 1–390 (2013)
72. Golstein, S.J., Rottenberg, F., Horlin, F., De Doncker, P., Sarrazin, J.: Physical layer security in an OFDM time reversal SISO communication with imperfect channel state information. IEEE Access **10**, 26778–26794 (2022)
73. Gotz, M., Rapp, M., Dostert, K.: Power line channel characteristics and their effect on communication system design. IEEE Commun. Mag. **42**(4), 78–86 (2004)
74. Gradshteyn, I., Ryzhik, I.: Table of Integrals, Series, and Products. Academic, Cambridge (2014)
75. Group, P.A.T.W., et al.: Draft specification for powerline intelligent metering evolution. Versión 1.3, **6** (2013)
76. Gunduz, M.Z., Das, R.: Cyber-security on smart grid: threats and potential solutions. Comput. Netw. **169**, 107094 (2020)
77. Güzelgöz, S., Celebi, H.B., Arslan, H.: Articulating factors defining RMS delay spread in LV PLC networks. J. Comput. Syst. Netw. Commun. **2010**, 1–9 (2010)
78. Güzelgöz, S., Çelebi, H.B., Arslan, H.: Statistical characterization of the paths in multipath PLC channels. IEEE Trans. Power Delivery **26**(1), 181–187 (2011)
79. Güzelgöz, S., Çelebi, H.B., Güzel, T., Arslan, H., Mıhçak, M.K.: Time frequency analysis of noise generated by electrical loads in PLC. In: 2010 17th International Conference on Telecommunications, pp. 864–871 (2010)

80. Hartmann, T., Fouquet, F., Klein, J., Le Traon, Y., Pelov, A., Toutain, L., Ropitault, T.: Generating realistic smart grid communication topologies based on real-data. In: 2014 IEEE International Conference on Smart Grid Communications (SmartGridComm), pp. 428–433 (2014)
81. Hashmat, R., Pagani, P., Zeddam, A., Chonavel, T.: MIMO communications for inhome PLC networks: Measurements and results up to 100 MHZ. In: ISPLC2010, pp. 120–124. IEEE, Piscataway (2010)
82. Hastings, J., Laverty, D.M., Morrow, D.J.: Securing the smart grid. In: 2014 49th International Universities Power Engineering Conference (UPEC), pp. 1–6. IEEE, Piscataway (2014)
83. HD-PLC Alliance: HD-PLC Alliance. http://www.hd-plc.org
84. He, H., Yan, J.: Cyber-physical attacks and defences in the smart grid: a survey. IET Cyb.-Phys. Syst. Theory Appl. **1**(1), 13–27 (2016)
85. Henkel, W., Graur, O.A., Islam, N.S., Pagel, U., Manak, N., Can, O.: Reciprocity for physical layer security with wireless FDD and in wireline communications. In: 2018 IEEE Globecom Workshops (GC Wkshps), pp. 1–6 (2018)
86. Henkel, W., Turjman, A.M., Kim, H., Qanadilo, H.K.H.: Common randomness for physical-layer key generation in power-line transmission. In: 2020 IEEE International Conference on Communications (ICC), pp. 1–6 (2020)
87. Hernandez Fernandez, J., Omri, A., Di Pietro, R.: Channel impulse response multilevel quantization for power line communications. IEEE Access **10**, 66113–66126 (2022)
88. Hernandez Fernandez, J., Omri, A., Di Pietro, R.: PLC physical layer link identification with imperfect channel state information. Energies **15**(16) (2022)
89. Hernandez Fernandez, J., Omri, A., Di Pietro, R.: Power grid surveillance: topology change detection system using power line communications. Int. J. Electr. Power Energy Syst. **145**, 108634 (2022)
90. Hernandez Fernandez, J., Wang, B., Massoud, A., Talib, S., Shhaab, M., Mohamed, A.: On ambient temperature of transformer substations in desert climates. In: 2021 IEEE Power Energy Society Innovative Smart Grid Technologies Conference (ISGT), pp. 1–4 (2021)
91. HomePlug Powerline Alliance: Homeplug 1.0: Technology white paper (2001)
92. HomePlug Powerline Alliance: Homeplug av white paper (2005)
93. HomePlug Powerline Alliance: Home plug green phy white paper: The standard for in-home smart grid powerline communications (2010)
94. HomePlug Powerline Alliance: Homeplug av2: Technology (2013)
95. Hong, J., Liu, C.C., Govindarasu, M.: Integrated anomaly detection for cyber security of the substations. IEEE Trans. Smart Grid **5**(4), 1643–1653 (2014)
96. Hota, A.R., Juvvanapudi, M., Bajpai, P.: Issues and solution approaches in PHEV integration to smart grid. Renew. Sustain. Energy Rev. **30**, 217–229 (2014)
97. Hubbard, B., Karasz, P., Reed, S.: Two major saudi oil installations hit by drone strike, and US blames Iran. The New York Times (Sept. 14, 2019), available at https://www.nytimes.com/2019/09/14/world/middleeast/saudi-arabia-refineries-drone-attack.html (Last Updated Sept 15, 2019) (2019)
98. Huo, Y., Prasad, G., Atanackovic, L., Lampe, L., Leung, V.C.: Grid surveillance and diagnostics using power line communications. 2018 IEEE International Symposium on Power Line Communications and its Applications, ISPLC 2018 pp. 1–6 (2018)
99. Huo, Y., Prasad, G., Lampe, L., Leung, V.C.: Cable health monitoring in distribution networks using power line communications. In: 2018 IEEE International Conference on Communications, Control, and Computing Technologies for Smart Grids, SmartGridComm 2018, pp. 1–6 (2018)
100. Huo, Y., Prasad, G., Atanackovic, L., Lampe, L., Leung, V.C.M.: Cable diagnostics with power line modems for smart grid monitoring. IEEE Access **7**, 60206–60220 (2019)
101. Huo, Y., Prasad, G., Lampe, L., Leung, V.C.M.: Advanced smart grid monitoring: Intelligent cable diagnostics using neural networks. In: 2020 IEEE International Symposium on Power Line Communications and its Applications (ISPLC), pp. 1–6 (2020)

102. Huo, Y., Prasad, G., Lampe, L., Leung, V.: Power Line Communication and Sensing Using Time Series Forecasting. Sensors **22**(14) (2022)
103. Hussain, S., Hernandez Fernandez, J., Al-Ali, A.K., Shikfa, A.: Vulnerabilities and countermeasures in electrical substations. Int. J. Crit. Infrastruct. Protect. **33**, 100406 (2021)
104. IEEE: IEEE std p1901.3: IEEE draft standard for broadband over power line networks: Medium access control and physical layer specifications amendment: Enhancement for internet of things applications
105. IEEE: IEEE 1901.2–2013 - IEEE standard for low-frequency (less than 500 khz) narrowband power line communications for smart grid applications (2010)
106. IEEE: IEEE 1901.1-2018 - IEEE standard for medium frequency (less than 12 mhz) power line communications for smart grid applications (2016)
107. IEEE: IEEE std p1901.1: IEEE draft standard for medium frequency (less than 15 mhz) power line communications for smart grid applications (2018)
108. Ikpehai, A., Adebisi, B., Rabie, K.M.: Broadband plc for clustered advanced metering infrastructure (AMI) architecture. Energies **9**(7), 569 (2016)
109. Irfan, M., Omri, A., Hernandez Fernandez, J., Oligeri, G.: Jamming detection in power line communications leveraging deep learning techniques. In: 2023 IEEE International Symposium on Networks, Computers and Communications (ISNCC) (2023)
110. ITU-T: Unified high speed wireline based home networking transceivers – power spectral density specification (2011)
111. ITU-T: Recommendation itu-t g.9904; narrowband orthogonal frequency division multiplexing power line communication transceivers for prime networks. Technical Report, ITU-T (2013)
112. ITU-T: Recommendation ITU-T g.9955; narrowband orthogonal frequency division multiplexing power line communication transceivers – physical layer specification (2014)
113. ITU-T: Recommendation ITU-T g.9956; narrowband orthogonal frequency division multiplexing power line communication transceivers – data link layer specification (2014)
114. ITU-T: Recommendation ITU-T g.9903; narrowband orthogonal frequency division multiplexing power line communication transceivers for g3-PLC networks. Technical Report, ITU-T (2017)
115. ITU-T: Recommendation ITU-T g.9960; unified high-speed wire-line based home networking transceivers - foundation (2019)
116. ITU-T: Recommendation ITU-T g.9961; unified high-speed wire-line based home networking transceivers – data link layer specification (2019)
117. ITU-T: Recommendation ITU-T g.9963; unified high-speed wireline-based home networking transceivers – multiple input/multiple output specification (2019)
118. Jarouf, A., Fernandez, J.H., Omri, A., Pietro, R.D.: Inferring Power Grid Information with Power Line Communications: Review and Insights. arXiv (2308.10598) (2023)
119. Jordan, T., Taylor, P.: Hacktivism and Cyberwars: Rebels with a Cause? Routledge, Milton Park (2004)
120. Kabalci, E., Kabalci, Y.: Smart Grids and Their Communication Systems. Springer, Berlin (2019)
121. Kaplan, S.M.: Smart Grid: Modernizing Electric Power Transmission and Distribution; Energy Independence, Storage and Security; Energy Independence and Security Act of 2007 (EISA); Improving Electrical Grid Efficiency, Communication, Reliability, and Resiliency; Integrating New and Renewable Energy Sources. The Capitol Net Inc (2009)
122. Karri, R., Rajendran, J., Rosenfeld, K., Tehranipoor, M.: Trustworthy hardware: identifying and classifying hardware trojans. Computer **43**(10), 39–46 (2010)
123. Kermani, M.M., Zhang, M., Raghunathan, A., Jha, N.K.: Emerging frontiers in embedded security. In: 2013 26th International Conference on VLSI Design and 2013 12th International Conference on Embedded Systems, pp. 203–208. IEEE, Piscataway (2013)
124. Knights, M.: Infrastructure targeting and postwar iraq. Policy Watch **0**(725) (2003)
125. Krekel, B., Adams, P., Bakos, G.: Occupying the information high ground: Chinese capabilities for computer network operations and cyber espionage. Int. J. Comput. Res. **21**(4), 333 (2014)

126. Kreps, S.E.: The 2006 lebanon war: lessons learned. Parameters **37**(1), 72 (2007)
127. Kumar, A., Agarwal, A.: Research issues related to cryptography algorithms and key generation for smart grid: A survey. In: 2016 7th India International Conference on Power Electronics (IICPE), pp. 1–5 (2016)
128. Kurt, M.N., Yılmaz, Y., Wang, X.: Real-time detection of hybrid and stealthy cyber-attacks in smart grid. IEEE Trans. Inf. Forens. Secur. **14**(2), 498–513 (2018)
129. Lampe, L., Ahmed, M.O.: Power grid topology inference using power line communications. In: 2013 IEEE International Conference on Smart Grid Communications (SmartGridComm), pp. 336–341 (2013)
130. Lampe, L., Tonello, A.M., Swart, T.: Power Line Communications: Principles, Standards and Applications from Multimedia to Smart Grid. Wiley, Hoboken (2016)
131. Lazar, S., Fisch, J., Reinert, C., Abraham, C.: Communication protocol over power line communication networks (2005). US Patent App. 10/666,852
132. Lee, C.B., Roedel, C., Silenok, E.: Detection and characterization of port scan attacks. Univeristy of California, Department of Computer Science and Engineering (2003)
133. Lee, K., Klingensmith, N., Banerjee, S., Kim, Y.: VoltKey: continuous secret key generation based on power line noise for zero-involvement pairing and authentication. Proc. ACM Interact. Mobile Wearable Ubiquit. Technol. **3**(3), 1–26 (2019)
134. Lehmann, A.M., Raab, K., Gruber, F., Fischer, E., Muller, R., Huber, J.B.: A diagnostic method for power line networks by channel estimation of PLC devices. In: 2016 IEEE International Conference on Smart Grid Communications, SmartGridComm 2016, pp. 320–325 (2016)
135. Lewis, J.A.: Assessing the Risks of Cyber Terrorism, Cyber War and Other Cyber Threats. Center Strategic & International Studies Washington, DC (2002)
136. Lewis, J., Baker, S.: The economic impact of cybercrime and cyber espionage. McAfee (2013)
137. Liang, G., Weller, S.R., Zhao, J., Luo, F., Dong, Z.Y.: The 2015 Ukraine blackout: Implications for false data injection attacks. IEEE Trans. Power Syst. **32**(4), 3317–3318 (2016)
138. Liang, D., Guo, H., Zheng, T.: Real-time impedance estimation for power line communication. IEEE Access **7**, 88107–88115 (2019)
139. Libicki, M.C.: Correlations between cyberspace attacks and kinetic attacks. In: 2020 12th International Conference on Cyber Conflict (CyCon), vol. 1300, pp. 199–213. IEEE, Piscataway (2020)
140. Lisowski, M., Masnicki, R., Mindykowski, J.: PLC-enabled low voltage distribution network topology monitoring. IEEE Trans. Smart Grid **10**(6), 6436–6448 (2019)
141. Liu, Y., Ning, P., Reiter, M.K.: False data injection attacks against state estimation in electric power grids. ACM Trans. Inf. Syst. Secur. **14**(1), 13 (2011)
142. Liu, Y., Chen, H., Wang, L.: Physical layer security for next generation wireless networks: theories, technologies, and challenges. IEEE Commun. Surveys Tutor. **19**(1), 347–376 (2017)
143. Ma, J.T.K., Liu, T.M., Wu, L.F.: New energy management system architectural design and intranet/internet applications to power systems. In: Proceedings of EMPD'98. 1998 International Conference on Energy Management and Power Delivery (Cat. No. 98EX137), vol. 1, pp. 207–212. IEEE, Piscataway (1998)
144. Maenou, T., Katayama, M.: Study on signal attenuation characteristics in power line communications. In: 2006 IEEE International Symposium on Power Line Communications and Its Applications, pp. 217–221 (2006)
145. Majumder, A., Caffery J, J.: Power line communications. IEEE Potentials **23**(4), 4–8 (2004)
146. Manchester, S., Mentz, J.A., Aftab, Y., Melder, A.H., Gangitano, G.J.: Configuring network settings for a power line networking device (2012). US Patent 8,270,934
147. Masood, R., Anwar, Z., et al.: Swam: Stuxnet worm analysis in metasploit. In: 2011 Frontiers of Information Technology, pp. 142–147. IEEE, Piscataway (2011)
148. Masood, B., Baig, S.: Standardization and deployment scenario of next generation NB-PLC technologies. Renew. Sustain. Energy Rev. **65**, 1033–1047 (2016)
149. Masood, B., Nazar, W., Masood, R.: Channel modeling of low voltage NB-PLC network using statistical and deterministic channel modeling approaches. In: 2018 7th International Conference on Renewable Energy Research and Applications (ICRERA), pp. 693–696 (2018)

150. Masood, B., Khan, M.A., Baig, S., Song, G., Rehman, A.U., Rehman, S.U., Asif, R.M., Rasheed, M.B.: Investigation of deterministic, statistical and parametric NB-PLC channel modeling techniques for advanced metering infrastructure. Energies **13**(12) (2020)
151. Mathur, A., Bhatnagar, M.R., Panigrahi, B.K.: PLC performance analysis over rayleigh fading channel under Nakagami-*m* additive noise. IEEE Commun. Lett. **18**(12), 2101–2104 (2014)
152. McGraw, G.: Cyber war is inevitable (unless we build security in). J. Strat. Stud. **36**(1), 109–119 (2013)
153. Melki, R., Noura, H.N., Hernandez Fernandez, J., Chehab, A.: Message authentication algorithm for OFDM communication systems. Telecommun. Syst. **76**, 1–20 (2020)
154. Meng, H., Chen, S., Guan, Y., Law, C., So, P., Gunawan, E., Lie, T.: A transmission line model for high-frequency power line communication channel. In: Proceedings. International Conference on Power System Technology, vol. 2, pp. 1290–1295 (2002)
155. Meng, H., Guan, Y., Chen, S.: Modeling and analysis of noise effects on broadband power-line communications. IEEE Trans. Power Delivery **20**(2), 630–637 (2005)
156. Milioudis, A.N., Andreou, G.T., Labridis, D.P.: Enhanced protection scheme for smart grids using power line communications techniques-part II: location of high impedance fault position. IEEE Trans. Smart Grid **3**(4), 1631–1640 (2012)
157. Milioudis, A.N., Andreou, G.T., Labridis, D.P.: Detection and location of high impedance faults in multiconductor overhead distribution lines using power line communication devices. IEEE Trans. Smart Grid **6**(2), 894–902 (2015)
158. Mlynek, P., Misurec, J., Koutny, M.: Random channel generator for indoor power line communication. Measure. Sci. Rev. **13**(4), 206–213 (2013)
159. Mlynek, P., Koutny, M., Misurec, J., Kolka, Z.: Measurements and evaluation of plc modem with g3 and prime standards for street lighting control. In: 18th IEEE International Symposium on Power Line Communications and Its Applications, pp. 238–243. IEEE, Piscataway (2014)
160. Mohan, V., Mathur, A.: Secrecy analysis of DCSK-based PLC systems with multiple eavesdroppers. IEEE Syst. J. **17**, 1–12 (2022)
161. Mohan, V., Mathur, A., Kaddoum, G.: Analyzing physical-layer security of PLC systems using DCSK: a copula-based approach. IEEE Open J. Commun. Soc. **4**, 104–117 (2023)
162. Mohsenian-Rad, A.H., Leon-Garcia, A.: Optimal residential load control with price prediction in real-time electricity pricing environments. IEEE Trans. Smart Grid **1**(2), 120–133 (2010)
163. Möller, D.P.F., Vakilzadian, H.: Ubiquitous networks: power line communication and internet of things in smart home environments. In: IEEE International Conference on Electro/Information Technology, pp. 596–601 (2014)
164. Moreno, J.A., Quintanilla, R.: Smartgrid applications using narrow band power line carrier in underground power distribution systems. PLC fault locator. In: CIRED 2009 - 20th International Conference and Exhibition on Electricity Distribution - Part 1, pp. 1–4 (2009)
165. Mukherjee, A.: Physical-layer security in the internet of things: sensing and communication confidentiality under resource constraints. Proc. IEEE **103**(10), 1747–1761 (2015)
166. Myerson, J.M.: Identifying enterprise network vulnerabilities. Int. J. Netw. Manag. **12**(3), 135–144 (2002)
167. Nakashima, E.: US target of massive cyber-espionage campaign. Washington post (2013)
168. Newman, R., Gavette, S., Yonge, L., Anderson, R.: Protecting domestic power-line communications. In: Proceedings of the Second Symposium on Usable Privacy and Security, pp. 122–132 (2006)
169. Noura, H.N., Melki, R., Chehab, A., Hernandez Fernandez, J.: Efficient and robust data availability solution for hybrid PLC/RF systems. Comput. Netw. **185**, 107675 (2020)
170. Ochôa, I.S., Leithardt, V.R., Zeferino, C.A., Silva, J.S.: Data transmission performance analysis with smart grid protocol and cryptography algorithms. In: 2018 13th IEEE International Conference on Industry Applications (INDUSCON), pp. 482–486. IEEE, Piscataway (2018)
171. Oksman, V., Galli, S.: G. hn: The new ITU-T home networking standard. IEEE Commun. Mag. **47**(10), 138–145 (2009)

172. Oksman, V., Zhang, J.: G. hnem: the new ITU-T standard on narrowband PLC technology. IEEE Commun. Mag. **49**(12), 36–44 (2011)
173. Oligeri, G., Sciancalepore, S., Raponi, S., Di Pietro, R.: Past-AI: Physical-layer authentication of satellite transmitters via deep learning. IEEE Trans. Inf. Forens. Secur. **18**, 274–289 (2023)
174. Omri, A., Hasna, M.O., Letaief, K.B.: Inter-relay interference management schemes for wireless multi-user decode-and-forward relay networks. IEEE Trans. Wirel. Commun. **14**(4), 2072–2081 (2015)
175. Omri, A., Hasna, M.O.: Modeling and performance analysis of D2D communications with interference management in 3-D HetNets. In: 2016 IEEE Global Communications Conference (GLOBECOM), pp. 1–7 (2016)
176. Omri, A., Hasna, M.O.: Physical layer security analysis of UAV based communication networks. In: 2018 IEEE 88th Vehicular Technology Conference (VTC-Fall), pp. 1–6 (2018)
177. Omri, A., Hernandez Fernandez, J., Sanz, A., Fliss, M.R.: PLC channel selection schemes for OFDM-based NB-PLC systems. In: 2020 IEEE International Symposium on Power Line Communications and its Applications (ISPLC), pp. 1–6 (2020)
178. Omri, A., Fernandez, J.H., Di Pietro, R.: Secure and successful transmission probability analysis for PLC networks. In: 2023 International Symposium on Networks, Computers and Communications (ISNCC), pp. 1–7 (2023)
179. Ottis, R.: Analysis of the 2007 cyber attacks against estonia from the information warfare perspective. In: Proceedings of the 7th European Conference on Information Warfare, p. 163. Academic Publishing Limited Reading, MA (2008)
180. Otuoze, A.O., Mustafa, M.W., Larik, R.M.: Smart grids security challenges: classification by sources of threats. J. Electr. Syst. Inf. Technol. **5**(3), 468–483 (2018)
181. Panasonic: Next-generation plc technology approved as baseline specification by IEEE p1901.3 working group. https://news.panasonic.com/global/press/data/2018/06/en180629-2/en180629-2.html
182. Pappu, S.J., Bhatt, N., Pasumarthy, R., Rajeswaran, A.: Identifying topology of low voltage distribution networks based on smart meter data. IEEE Trans. Smart Grid **9**(5), 5113–5122 (2018)
183. Passerini, F., Tonello, A.M.: On the exploitation of admittance measurements for wired network topology derivation. IEEE Trans. Instrument. Measure. **66**(3), 374–382 (2017)
184. Passerini, F., Tonello, A.M.: Smart grid monitoring using power line modems: anomaly detection and localization. IEEE Trans. Smart Grid **10**(6), 6178–6186 (2019)
185. Passerini, F., Tonello, A.M.: Smart grid monitoring using power line modems: effect of anomalies on signal propagation. IEEE Access **7**, 27302–27312 (2019)
186. Passerini, F., Tonello, A.M.: Secure PHY layer key generation in the asymmetric power line communication channel. Electronics **9**(4), 605 (2020)
187. Pavlidou, N., Vinck, A.H., Yazdani, J., Honary, B.: Power line communications: state of the art and future trends. IEEE Commun. Mag. **41**(4), 34–40 (2003)
188. Peppanen, J., Grijalva, S., Reno, M.J., Broderick, R.J.: Distribution system low-voltage circuit topology estimation using smart metering data. In: 2016 IEEE/PES Transmission and Distribution Conference and Exposition (T D), pp. 1–5 (2016)
189. Picchi, D., Lourenço, M., da Silva, A., Nascimento, D., Saldanha, E., Dantas, I., Resende, J.: Learnings from pilot implementation of smart city by a brazilian energy utility. In: Smart Cities Technologies. IntechOpen (2016)
190. Pinto-Benel, F.A., Cruz-Roldán, F.: 2-ASCET for broadband multicarrier transmission over in-home and in-vehicle power line networks. In: 2015 IEEE 18th International Conference on Intelligent Transportation Systems, pp. 1351–1356 (2015)
191. Pittolo, A., Tonello, A.M.: Physical layer security in PLC networks: achievable secrecy rate and channel effects. In: 2013 IEEE 17th International Symposium on Power Line Communications and Its Applications, pp. 273–278 (2013)
192. Pittolo, A., Tonello, A.M.: Physical layer security in power line communication networks: an emerging scenario, other than wireless. IET Commun. **8**(8), 1239–1247 (2014)

193. Pittolo, A., Tonello, A.M.: Physical layer security in power line communication networks. In: Baldi, M., Tomasin, S. (eds.) Physical and Data-Link Security Techniques for Future Communication Systems, pp. 125–144. Springer International Publishing, Cham (2016)
194. Pittolo, A., Tonello, A.M.: Physical layer security in power line communication networks. Lect. Notes Electr. Eng. Book Ser. **358**, 125–144 (2016)
195. Prasad, G., Huo, Y., Lampe, L., Leung, V.C.M.: Machine learning based physical-layer intrusion detection and location for the smart grid. In: 2019 IEEE International Conference on Communications, Control, and Computing Technologies for Smart Grids (SmartGridComm), pp. 1–6 (2019)
196. Prasad, G., Taghizadeh, O., Lampe, L., Mathar, R.: Securing MIMO power line communications with full-duplex jamming receivers. In: 2019 IEEE International Symposium on Power Line Communications and its Applications (ISPLC), pp. 1–6 (2019)
197. PRIME Alliance: PRIME Alliance - interoperable standard for advance meter management and smart grid. https://www.prime-alliance.org
198. PRIME Alliance: PRIME broadband powerline. https://www.prime-alliance.org/solutions/prime-bpl
199. PRIME Alliance: Draft specification for powerline intelligent metering evolution, r1.3.6. Technical Report, PRIME Alliance (2013)
200. PRIME Alliance: Prime v1.4 white paper. Technical Report, PRIME Alliance (2014). https://www.prime-alliance.org/wp-content/uploads/2014/10/whitePaperPrimeV1p4_final.pdf
201. PRIME Alliance: Specification for powerline intelligent metering evolution, r1.4. Technical Report, PRIME Alliance (2014)
202. Rahman, M.M., Hong, C.S., Lee, S., Lee, J., Razzaque, M.A., Kim, J.H.: Medium access control for power line communications: an overview of the IEEE 1901 and ITU-T g. hn standards. IEEE Commun. Mag. **49**(6), 183–191 (2011)
203. Rao, R., Akella, S., Guley, G.: Power line carrier (PLC) signal analysis of smart meters for outlier detection. In: 2011 IEEE International Conference on Smart Grid Communications, SmartGridComm 2011, pp. 291–296 (2011)
204. Raponi, S., Fernandez, J.H., Omri, A., Oligeri, G.: Long-term noise characterization of narrowband power line communications. IEEE Trans. Power Delivery **37**(1), 365–373 (2022)
205. Razazian, K., Umari, M., Kamalizad, A., Loginov, V., Navid, M.: G3-PLC specification for powerline communication: Overview, system simulation and field trial results. In: ISPLC2010, pp. 313–318. IEEE, Piscataway (2010)
206. Rihan, M.: Applications and requirements of smart grid. In: Smart Grids and Their Communication Systems, pp. 47–79. Springer Singapore, Singapore (2019)
207. Romero Agüero, J., Takayesu, E., Novosel, D., Masiello, R.: Grid modernization: challenges and opportunities. Electr. J. **30**(4), 1–6 (2017)
208. Russakovsky, O., et al.: ImageNet large scale visual recognition challenge. Int. J. Comp. Vision **115**(3), 211–252 (2015)
209. Salem, A., Rabie, K.M., Hamdi, K.A., Alsusa, E., Tonello, A.M.: Physical layer security of cooperative relaying power-line communication systems. In: 2016 International Symposium on Power Line Communications and Its Applications (ISPLC), pp. 185–189. IEEE, Piscataway (2016)
210. Salem, A., Hamdi, K.A., Alsusa, E.: Physical layer security over correlated log-normal cooperative power line communication channels. IEEE Access **5**, 13909–13921 (2017)
211. Sánchez-Martínez, J.J., Cortés, J.A., Díez, L., Cañete, F.J., Torres, L.M.: Performance analysis of OFDM modulation on indoor plc channels in the frequency band up to 210 MHZ. In: ISPLC2010, pp. 38–43 (2010)
212. Sankar, L., Rajagopalan, S.R., Mohajer, S.: Smart meter privacy: a theoretical framework. IEEE Trans. Smart Grid **4**(2), 837–846 (2013)
213. Schmitt, M.N.: Tallinn Manual on the International Law Applicable to Cyber Warfare. Cambridge University Press, Cambridge (2013)
214. Schwager, A., Stadelmeier, L., Zumkeller, M.: Potential of broadband power line home networking. In: Second IEEE Consumer Communications and Networking Conference, 2005. CCNC. 2005, pp. 359–363 (2005)

215. Sharma, K., Saini, L.M.: Power-line communications for smart grid: Progress, challenges, opportunities and status. Renew. Sustain. Energy Rev. **67**, 704–751 (2017)
216. Sheffi, Y.: Supply chain management under the threat of international terrorism. Int. J. Logist. Manag. **12**(2), 1–11 (2001)
217. Shin, I., Cho, M.: On localized countermeasure against reactive jamming attacks in smart grid wireless mesh networks. Appl. Sci. **8**(12), 2340 (2018)
218. Slacik, J., Mlynek, P., Rusz, M., Musil, P., Benesl, L., Ptacek, M.: Broadband power line communication for integration of energy sensors within a smart city ecosystem. Sensors **21**(10), 3402 (2021)
219. Smail, M.K., Pichon, L., Olivas, M., Auzanneau, F., Lambert, M.: Recent progress in emc and reliability for automotive applications. In: VXV International Symposium on Theoretical Engineering, pp. 1–5 (2009)
220. Smail, M.K., Pichon, L., Olivas, M., Auzanneau, F., Lambert, M.: Detection of defects in wiring networks using time domain reflectometry. IEEE Trans. Mag. **46**(8), 2998–3001 (2010)
221. Smart Energy International: PRIME addresses new communication challenges embracing BPL. https://www.smart-energy.com/industry-sectors/smart-meters/prime-addresses-new-communication-challenges-embracing-bpl
222. Smart Metering International: Maxim and ERDF release the first OFDM-based powerline-communication specification supporting the ipv6 internet protocol for smart grids (2009). https://www.smart-energy.com/regional-news/north-america/first-ofdm-based-powerline-communication-specification-released/
223. Stoyanov, I., Iliev, T., Mihaylov, G., Ivanova, E., Kogias, P.: Smart grid communication protocols in intelligent service for household energy use. In: Silhavy, R., Silhavy, P., Prokopova, Z. (eds.) Cybernetics Approaches in Intelligent Systems, pp. 380–389. Springer International Publishing, Cham (2018)
224. Sullivan, J.E., Kamensky, D.: How cyber-attacks in ukraine show the vulnerability of the us power grid. Electr. J. **30**(3), 30–35 (2017)
225. Sunguk, L.: Security issues of power line multi-home networks for seamless data transmission. Int. J. Adv. Sci. Technol. **37** (2011)
226. Takmaz, E.: Impedance, attenuation and noise measurements for power line communication. In: 2016 4th International Istanbul Smart Grid Congress and Fair (ICSG), pp. 1–4 (2016)
227. Tan, S., De, D., Song, W.Z., Yang, J., Das, S.K.: Survey of security advances in smart grid: a data driven approach. IEEE Commun. Surveys Tutor. **19**(1), 397–422 (2017)
228. Thonnard, O., Bilge, L., O'Gorman, G., Kiernan, S., Lee, M.: Industrial espionage and targeted attacks: Understanding the characteristics of an escalating threat. In: International Workshop on Recent Advances in Intrusion Detection, pp. 64–85. Springer, Berlin (2012)
229. Treytl, A., Novak, T.: Practical issues on key distribution in power line networks. In: 2005 IEEE Conference on Emerging Technologies and Factory Automation, vol. 2, pp. 8–90 (2005)
230. Tsokalo, I., Lehnert, R.: Modeling approach of broadband in-home PLC in network simulator 3. In: 2015 IEEE International Symposium on Power Line Communications and Its Applications (ISPLC), pp. 113–118. IEEE, Piscataway (2015)
231. Tweneboah-Koduah, S., Tsetse, A.K., Azasoo, J., Endicott-Popovsky, B.: Evaluation of cybersecurity threats on smart metering system. In: Information Technology-New Generations, pp. 199–207. Springer, Berlin (2018)
232. Ulaby, F., Ravaioli, U.: Fundamentals of Applied Electromagnetics. Pearson Education, London (2015)
233. Uribe-Pérez, N., Hernández, L., Gómez, R., Soria, S., de la Vega, D., Angulo, I., Arzuaga, T., Gutiérrez, L.: Smart management of a distributed generation microgrid through PLC prime technology. In: 2015 International Symposium on Smart Electric Distribution Systems and Technologies (EDST), pp. 374–379. IEEE, Piscataway (2015)
234. Uribe-Pérez, N., Angulo, I., de la Vega, D., Arzuaga, T., Fernández, I., Arrinda, A.: Smart grid applications for a practical implementation of IP over narrowband power line communications. Energies **10**(11), 1782 (2017)

235. Usman, A., Shami, S.H.: Evolution of communication technologies for smart grid applications. Renew. Sustain. Energy Rev. **19**, 191–199 (2013)
236. Valocký, F., Puchalik, M., Orgon, M.: Implementing asymmetric cryptography in high-speed data transmission over power line. In: 2020 11th IEEE Annual Ubiquitous Computing, Electronics & Mobile Communication Conference (UEMCON), pp. 0849–0854 (2020)
237. Vigo, R., Yüksel, E., Ramli, C.D.P.K.: Smart grid security a smart meter-centric perspective. In: 2012 20th Telecommunications Forum (TELFOR), pp. 127–130. IEEE, Piscataway (2012)
238. Waadt, A., Kocks, C., Bruck, G.H., Jung, P., Sachsenhauser, B.: Achievable data rates of broadband power line communications in an underground medium-voltage network. J. Green Eng. **3**(3), 245–259 (2013)
239. Wang, T., Liu, Y., Athanasios, A.: Survey on channel reciprocity based key establishment techniques for wireless systems. Wirel. Netw. **21**(6), 1835–1846 (2015)
240. Wang, X., Dai, X., Xian, H., Song, W., Gao, B., Xu, S., Pang, L., Guo, S.: Fault localization for power line communications with topology inference. In: Proceedings - 2019 International Conference on Intelligent Computing, Automation and Systems, ICICAS 2019, pp. 85–90 (2019)
241. Weers, D., Shamsedin, M.: Testing a new direct load control power line communication system. IEEE Trans. Power Delivery **2**(3), 657–660 (1987)
242. Weinthal, E., Sowers, J.: Targeting infrastructure and livelihoods in the west bank and gaza. Int. Affairs **95**(2), 319–340 (2019)
243. Weissbrodt, D.: Cyber-conflict, cyber-crime, and cyber-espionage. Minn. J. Int'l L. **22**, 347 (2013)
244. White, S.P.: Understanding Cyberwarfare: Lessons from the Russia-Georgia War. Modern War Institute at West Point (2018)
245. Whyte, A., Whittaker, M.: DS2 introduces 400 mbps powerline communications technology. https://www.targetwire.com/targetwire/2007/10/18/tw396/tw396_uk.html (2007)
246. Wueest, C.: Targeted attacks against the energy sector. Symantec Security Response, Mountain View, CA (2014)
247. Yaacoub, J.P., Salman, O.: Security analysis of drones systems: attacks, limitations, and recommendations. Int. Things **11**, 100218 (2020)
248. Yaacoub, J.P.A., Noura, M., Noura, H.N., Salman, O., Yaacoub, E., Couturier, R., Chehab, A.: Securing internet of medical things systems: limitations, issues and recommendations. Fut. Gener. Comput. Syst. **105**, 581–606 (2020)
249. Yaacoub, J.P.A., Salman, O., Noura, H.N., Kaaniche, N., Chehab, A., Malli, M.: Cyber-physical systems security: limitations, issues and future trends. Microprocess. Microsyst. **77**, 103201 (2020)
250. Yaacoub, J.P., Hernandez Fernandez, J., Noura, H.N., Chehab, A.: Security of power line communication systems: issues, limitations and existing solutions. Comput. Sci. Rev. **39**, 100331 (2021)
251. Yan, Y., Qian, Y., Sharif, H., Tipper, D.: A survey on smart grid communication infrastructures: motivations, requirements and challenges. IEEE Commun. Surveys Tutor. **15**(1), 5–20 (2013)
252. Yang, F., Ding, W., Song, J.: Non-intrusive power line quality monitoring based on power line communications. In: 2013 IEEE 17th International Symposium on Power Line Communications and Its Applications, pp. 191–196 (2013)
253. Yang, F., Islam, M.A., Ren, S.: PowerKey: Generating secret keys from power line electromagnetic interferences. In: International Conference on Network and System Security, pp. 354–370. Springer, Berlin (2020)
254. Yonge, L., Abad, J., Afkhamie, K., Guerrieri, L., Katar, S., Lioe, H., Riva, R., Schneider, D., Schwager, A.: An overview of the homeplug av2 technology. J. Electr. Comput. Eng. **2013**, 1–20 (2013)
255. Yoon, Y.J., Choi, J.W., Kim, S.C., Lee, J.H., Kim, Y.H.: The optimal power allocation for security in multicarrier relay power line communication. In: 2016 International Symposium on Power Line Communications and its Applications (ISPLC), pp. 190–195 (2016)

256. Yousuf, M.S., El-Shafei, M.: Power line communications: An overview - Part I. In: 2007 Innovations in Information Technologies (IIT), pp. 218–222 (2007)
257. Zhang, C., Zhu, X., Huang, Y., Liu, G.: High-resolution and low-complexity dynamic topology estimation for PLC networks assisted by impulsive noise source detection. IET Commun. **10**(4), 443–451 (2016)
258. Zhang, J., Duong, T.Q., Marshall, A., Woods, R.: Key generation from wireless channels: a review. IEEE Access **4**, 614–626 (2016)
259. Zhang, J., Liu, X., Cui, Y., Xu, D.: Physical-layer secret key generation in power line communication networks. IEEE Access **10**, 48539–48550 (2022)
260. Zhao, X., Qi, Y., Li, G.: Research and implementation of PLC for multiport traveling wave fault location in the medium voltage distribution network. In: 2011 4th International Conference on Electric Utility Deregulation and Restructuring and Power Technologies (DRPT), pp. 614–617 (2011)
261. Zhu, W., Zhu, X., Lim, E., Huang, Y.: State-of-art power line communications channel modelling. Procedia Comput. Sci. **17**, 563–570 (2013)
262. Zimmerman, M., Dostert, K.: Analysis and modeling of impulsive noise in broad-band powerline communications. IEEE Trans. Electromagn. Compatib. **44**(1), 249–258 (2002)
263. Zimmermann, M., Dostert, K.: A multipath model for the powerline channel. IEEE Trans. Commun. **50**(4), 553–559 (2002)

Index

A
Additive Noise in PLC, 28
Adversary Model, 114
Applications of PLC, 16
Association of Radio Industries and Businesses (ARIB), 9
Authentication, 35
Availability, 35
Average Secrecy Capacity, 53

B
Bit Mismatch Rate, 75
Botnet, 35
Broadband Power Line Communication (BB-PLC), 10
Brute-force/dictionary, 39
Buffer overflow attacks, 38

C
CENELEC Frequency Band, 11
Channel Impulse Response (CIR), 5
Channel Probing, 68
Channel State Information, 23
Channel Transfer Function (CTF), 23
China Electric Power Research Institute (CEPRI), 9
Classification of PLC applications, 19
Confidentiality, 33
Convolutional Neural Network (CNN), 114
Criminal groups, 37
Cyber-military operations, 35

D
Data Attacks, 40
Denial of Service (DoS), 114
Deterministic Modeling Approach, 66
Direct load control, 40

E
European Telecommunications Standards Institute (ETSI), 9

F
False Negative Detection Probability (FNDP), 108
False Positive Detection Probability (FPDP), 108
Federal Communications Commission (FCC), 9
Finite Impulse Response (FIR), 67

G
G3-PLC, 14
G.hn, 15
G.hn-MIMO, 15
G.hnem, 14

J. Hernandez Fernandez et al., *Physical Layer Security in Power Line Communications*, Advances in Information Security 108,
https://doi.org/10.1007/978-3-031-57349-1

Grid Cybersecurity, 20
Grid Monitoring, 18

H
High Definition-Power Line Communication (HD-PLC), 15
Home Area Networks (HAN), 10
HomePlug, 16

I
Identity Forgery, 39
IEEE P1901.3, 15
Indirect load control, 40
Institute of Electrical and Electronics Engineers (IEEE), 9
Integrity, 33
Interception, 39
International Telecommunication Union (ITU), 9

J
Jamming, 113
Jamming Detection Challenges, 113

L
LAN attacks, 39
Load alteration attacks, 40
Logic bombs, 38
Low-Frequency Narrow-Band Power Line Communications (LF SGPLC), 14

M
Malicious agents, 35
Malicious Applications, 38
Malware, 37
Medium Frequency Power Line Communications (MFSGPLC), 14
Misconfiguration, 38

N
Narrow band Power Line Communication (NB-PLC), 10
National Standard Developing Organizations (NDSO), 9
NB-PLC Technologies, 11

P
Phishers & spammers, 37
Physical Layer Security (PLS), 4
PLC for Data Transmission, 16
PLC for Grid Information Inference, 18
PLC Issues & Limitations, 31
PLC Multipath Channel Model, 101
PLC Multipath Propagation Model, 22
PLC Security Attacks, 39
PLC Security Threats, 33
PLC wiretap channel model, 116
PLS Key Generation Scheme, 68
Power Line Intelligent Metering Evolution (PRIME), 11
Power Line Modem (PLM), 18
Probability Distribution Function (PDF), 52

Q
Quantization, 68

R
Ransomware, 38
Reconciliation, 70, 75

S
Secrecy Outage Probability, 55
Secure and Successful Transmission Probability (SSTP), 57
Sniffing, 39
Spies, 35
Spyware, 38
Standard Development Organization (SDO), 9
Successful Detection Probability (SDP), 108
Successful Path Detection Gain (SPDG), 92
Successful Path Detection Probability (SPDP), 92

T
Terrorists, 37
Threat Types, 37
Topology Change Detection Algorithm, 106
Topology Simulation, 73
Transmission Line Theory, 21
Transmission Line (TL), 66
Transverse Electromagnetic (TEM), 21
Trojans, 38

U
Ultra Narrow band Power Line Communication (UNB-PLC), 10

V
VCIR Generation, 74
Viruses, 38

W
Web interface, 38
Worms, 38

Z
Zero-day exploit, 39

MIX
Papier aus verantwortungsvollen Quellen
Paper from responsible sources
FSC® C105338
FSC
www.fsc.org

If you have any concerns about our products,
you can contact us on
ProductSafety@springernature.com

In case Publisher is established outside the EU,
the EU authorized representative is:
Springer Nature Customer Service Center GmbH
Europaplatz 3, 69115 Heidelberg, Germany

Printed by Libri Plureos GmbH
in Hamburg, Germany